AF721637

The Explosion of Life Forms

SCIENCES

Biology, Field Director – Marie-Christine Maurel

Xenobiology, Exobiology, Astrobiology, Origins of Life, Subject Heads – Marie-Christine Maurel and Marc Ollivier

The Explosion of Life Forms

Living Beings and Morphology

Coordinated by
Georges Chapouthier
Marie-Christine Maurel

iSTE

WILEY

First published 2020 in Great Britain and the United States by ISTE Ltd and John Wiley & Sons, Inc.

Apart from any fair dealing for the purposes of research or private study, or criticism or review, as permitted under the Copyright, Designs and Patents Act 1988, this publication may only be reproduced, stored or transmitted, in any form or by any means, with the prior permission in writing of the publishers, or in the case of reprographic reproduction in accordance with the terms and licenses issued by the CLA. Enquiries concerning reproduction outside these terms should be sent to the publishers at the undermentioned address:

ISTE Ltd
27-37 St George's Road
London SW19 4EU
UK

www.iste.co.uk

John Wiley & Sons, Inc.
111 River Street
Hoboken, NJ 07030
USA

www.wiley.com

© ISTE Ltd 2020
The rights of Georges Chapouthier and Marie-Christine Maurel to be identified as the authors of this work have been asserted by them in accordance with the Copyright, Designs and Patents Act 1988.

Library of Congress Control Number: 2020937596

British Library Cataloguing-in-Publication Data
A CIP record for this book is available from the British Library
ISBN 978-1-78945-005-7

ERC code:
LS8 Ecology, Evolution and Environmental Biology
LS8_6 Phylogenetics, systematics, comparative biology
LS8_7 Macroevolution, paleobiology

MIX
Paper from responsible sources
FSC® C013604
www.fsc.org

Contents

Introduction

Georges CHAPOUTHIER[1] and Marie-Christine MAUREL[2]

[1] *ICM-CNRS, Paris, France*

[2] *ISYEB-CNRS-MNHN, Sorbonne University, Paris, France*

One of the essential characteristics of living beings, perhaps the most important, is an explosion of their forms, linked to the extreme diversity of environments to which they have been able to adapt. It is precisely the multiplicity of these astonishing morphologies, in all its facets, that this book aims to explore.

Of course, this diversity is rooted in the very origin of life itself, as Marie-Christine Maurel relates. The diversity of physical and chemical conditions on Earth during the Hadean period, as well as the diversity of planets and the multiple bodies of the solar system, may have led to the formation of molecules, organizations and protocells with a wide variety of shapes and metabolisms. The different chemical elements of the primitive ocean and fossil traces from the first stromatolites could provide evidence of this. Thus, biodiversity is original and current research is discovering multiple ways of "existing", far from a standardized vision of life.

These considerations lead, of course, to the great founding question: what is life? This is what Andreas Losch tries to answer. If it is clear that, in the final analysis, living matter can be summed up as creations of physics and chemistry, and if it is true that there is no such thing as a mysterious "vital principle", would life not then find its specificity precisely in the very explosion of its forms? But then a question of language arises. What we call "life" in everyday language, what we have an existential experience of, is not necessarily what life is at the scientific level, where material bases matter more than forms. As with other complex concepts, "life" is in fact an ill-defined combination of properties that mineral substances do not possess,

The Explosion of Life Forms,
coordinated by Georges CHAPOUTHIER and Marie-Christine MAUREL. © ISTE Ltd 2020.

where form plays a key role, and "words to describe it" are still missing. Indeed, how will "robots" and other machines describe it in a few decades?

In living beings that are already constituted, form becomes, as Jean-Pierre Gasc shows, a means of knowledge. From Aristotle to Cuvier, function is interpreted in relation to morphology. Nowadays, more emphasis is placed on the weight of the physical framework, which suggests that form is an adaptation to environmental constraints, as evidenced for example by the symmetry of bilaterians, the appearance of even appendages (allowing "'appendicular' movement on land") or the "cephalization" of mobile animals. In the second chapter, Jean-Pierre Gasc recalls the central role of the theses of the Scottish zoologist D'Arcy Wenworth Thompson, in his famous book *On Growth and Form*, which holds "an almost legendary place in scientific literature" and which makes him, in a way, the prophet of the link between form and functional adaptation.

We know today that, while genes define the major parameters of the form of living beings, the essential part, the definitive shape, is constituted by the almighty processes of epigenesis. Jonathan B. Weitzman leads us along these astonishing embryological paths which, going beyond the genetic basis, lead, in successive stages, to the generation of form in the developing embryo. Today, much is known about the epigenetic mechanisms by which environmental signals and genetic characteristics are integrated and contribute to the formation of beautiful forms, underlying the *epigenetic landscape* so dear to Waddington's heart.

The very original world of protists and bacteria is presented to us by Guillermo Paz-y-Miño-C and Avelina Espinosa. The numerous strategies by which these beings detect and communicate with each other to survive and develop offer striking examples of the explosion of forms at the microscopic level. These processes also provide a better understanding of the evolutionary strategies involved in the transformation of forms, such as inclusive selective value, whose adaptive success is due to direct descendants but also to kin selection.

Florian Jabbour and Guilhem Mansion tackle the fascinating field of plant forms. "The myriad of plant forms has always been – and still is – a source of astonishment, contemplation and even inspiration to which man cannot remain oblivious". However, the authors show that this diversity, which could be described as 'anthropocentric', is not the same one perceived by other representatives of the animal kingdom, such as pollinating insects, "often sensitive to signals that can only be observed in ultraviolet light". The authors analyze how plant forms emerge, both in terms of individual development (ontogenesis) and lineage evolution (phylogenesis), and how these evolutions allow plants to adapt to the surrounding environment and the constraints it imposes, as well as the possible constraints of internal operation. Lastly, the authors devote a large part of their chapter to all the ways in which human societies have used these plant forms for their own benefit.

Forms are not only manifested in the architecture of living beings, but also, particularly in animals, in their performances and achievements. Robert Jaffard addresses one of the most central performances of the psyche: memory. He analyzes the different forms of memory that allow an individual to adapt its behavior to the constraints and changes in its environment. Ranging from the simple phenomenon of habituation present in invertebrates to the highly sophisticated ability to form memories, these different forms of memory are juxtaposed as the nervous system becomes more complex.

Another area that contributes, in an essential way, to the performance of animals is that of sensory universes, of which Dalila Bovet offers us a detailed overview. Chemical sensitivities such as taste or smell, mechanical sensitivities such as touch or hearing, electromagnetic sensitivities such as vision, but also electrical, magnetic or thermal sensitivities; here are some of the innumerable sensory forms that have enabled animals to develop such particularly complex behaviors.

Beyond living beings, robots are built by human beings on basics inspired by living forms. In particular, they possess very interesting social and emotional forms, which Laurence Devillers explains. She shows that, although these forms are simulated, we "anthropomorphize" the behavior of machines. We imagine living things in artificial forms. She therefore explains the need to talk about ethics and transparency in the design of these machines that non-experts are far from understanding, and who could easily manipulate us.

On the technological level, mimicking the living, its forms, in order to develop and design medical technologies – orthoses, prostheses and other systems of electrophysiological or cognitive functional substitution, intra or extracorporeal – is a necessity in order to reconstruct anatomy and restore physiology. After placing this question in a historical and epistemological perspective, Didier Fass extends critical reflection to forms of thought or treatment by medical and symbiotic machines. He explains the associated concepts and the theoretical principles necessary to understand the organization of living organisms, which are essential for designing and modeling medical artifacts in integrative bioengineering. Thus, he describes the basic needs for biocompatibility and bio-integration as well as the fundamental requirements, including those of form(s), of "human-artificial" engineering for medicine and for "augmented humans". In conclusion, the author proposes a perspective of integrative bioengineering with the bioprinting of medical devices and tissue engineering.

On a more general and philosophical level, Georges Chapouthier wonders whether, despite their extreme diversity, biological forms (both anatomical and mental), and consequently man-made technological forms, which have a common origin, might not also have common principles of construction. According to him, all forms of life or systems that mimic life would indeed result from the repeated application of two main principles: *juxtaposition* of similar units; and then *integration* of these units to form more complex structures, of which the original units then become parts. Just as in a mosaic, the new complex structures (the "whole") then leave a degree of autonomy to the parts that make them up.

The book ends with a philosophical reflection by Jean-Michel Besnier on the very role of "form" in our 21st Century society. The current trend seems to be to forget form and replace it with a fluid potentiality where "nothing remains, everything evolves and offers itself to mutations and transformations". From this perspective, almighty "information" does not just aim to complement form. It aims, according to some authors, to totally replace it with "power", "a process that has gotten rid of formal conditions". Besnier analyzes and underlines the dangers of this process, a triumph of blind technology, supported by transhumanist, often catastrophist currents, where "power finally escapes the mind, that is to say, human control". Through this "deformation", "the absence of form is equivalent to violence and death".

This superb conclusion gives all its relevance and all its legitimacy to our work, which is, as we have seen, a plea, with multiple voices, for the living form "in all its forms", but also, as a consequence and by the same token, a profound plea for life itself.

1

Possible Traces and Clues of Early Life Forms

Marie-Christine MAUREL
ISYEB-CNRS-MNHN, Sorbonne University, Paris, France

"The pivotal question of the living ... is that of morphogenesis."
Claude Bernard, 1878

1.1. Introduction

Since its formation 4.5 billion years ago, the Earth has occupied a unique place in the solar system. The presence of liquid water, the first cradle of life, on 70% its surface is one of the major features of its appearance: 90% of life's history has taken place in water and we still ask ourselves numerous questions about its formation today.

Around 3.85 billion years ago, mineral matter and organic matter intertwined to produce, by chemical reactions, the amino acids of proteins, the constituents of the nucleic acids RNA and DNA (nitrogenous bases, sugars, phosphates), and all sorts of blocks that contributed to the design of the matrix of living things. Less than a billion years after the accretion of our planet, the elementary building blocks of biological molecules were thus formed on the primitive Earth. Today, life is omnipresent, covering the entire planet and its systems, at all latitudes, including our skin and our digestive tract, inhabited by thousands of bacteria species.

The Explosion of Life Forms,
coordinated by Georges CHAPOUTHIER and Marie-Christine MAUREL. © ISTE Ltd 2020.

There is a very wide variety of shapes, from rod-shaped bacilli to spherical shells, as well as helical, spiral or star-shaped structures, to name but a few of the forms observed in microscopic bacteria, that are living either in isolation, in filamentous association, in symbiosis, etc. The diversity of shapes and sizes is also observed in viral particles (Adriaenssens *et al.* 2018) and in multicellular organisms such as humans, snails, ferns, geckos, etc.

And diversity also manifests itself over time: nowadays, we no longer find pithecanthropes, lepidodendrons, tyrannosaurs, ammonite trilobites, etc.

Are the varieties observed today the only ones possible? Are there other paths, other formats, other modes that have not (yet) been explored by nature? Or by our own understanding? Or by our technological limits?

The total number of living species is estimated at 10^{12}, and only 10^5 of these have been identified to date (Locey and Lennon 2016). We only know 2–10% of the species that exist today, which represents 1/1,000 of the species that have existed for 3.85 billion years (Mora *et al.* 2011).

These data alone justify the weakness of our generalizations.

1.2. Have "things" always been as they are today?

What prebiotic chemical reactions can reasonably be simulated in the laboratory? The attributes of today's living organisms (structures, perennial and hereditary metabolic pathways) have passed through billions of years of planetary, geological and environmental events. Natural selection, by not retaining what was becoming inadequate, has amplified new attempts and taken the place of what was left vacant by recent extinctions, thus favoring innovation. This is how new species settle into diversity.

It is worth recalling that Jean-Baptiste Lamarck, in *Philosophie zoologique* (1809), was the first to develop a physical theory of living beings, that is to say, an organization of the subject driven by a series of physical processes. He considered that the simplest beings were formed in an appropriate environment in response to physical-chemical laws. According to Lamarck, the living beings resulting from this spontaneous generation adapted to the environment and thus became more "complex".

The invention of the microscope in the 21st Century revealed that all living organisms are made up of cells of different shapes. Since then, the cell has been regarded as the fundamental structural and functional biological unit of all living beings.

A student of Justus von Liebig, Moritz Traube, developed an "artificial mineral cell" from copper and potassium ferrocyanide in 1867. He observed the growth and budding of structures that resembled cell-like forms like those observed by Robert Hooke in 1665.

Then, Stéphane Leduc, inventor of the term "synthetic biology" (1912), declared at the beginning of the 20th Century, despite the craze for chemistry at the time: "Why is it less acceptable to try to find out how to make a cell than to make a molecule?"

Later, in the 1920s, the Soviet biochemist Alexander Oparin and the English geneticist John Burdon Sanderson Haldane proposed that elementary bricks of life, formed from gaseous elements in the atmosphere, were deposited in the primitive ocean, creating a "prebiotic soup" rich in assorted molecules that, when assembled, formed protocells or coacervates. From then on, all sorts of microspheres, micelles, protobionts and other models of single cells were imagined to represent simple compartments.

In order to "come alive", the first processes were therefore able to take place in micro-environments that were capable of keeping the different components close to each other, thus promoting their interactions. Initially, this could be in the hollow of a rock or on clay dust, then in tiny organic vesicles, a kind of small bag that fills with molecules until it splits in two.

Coacervates were synthesized in a laboratory in the 1930s by Bungenberg de Jong. Proteinoids (Fox 1988), microspheres, marigranules and other compartmentalizing structures were obtained in order to mimic early cell forms.

In 1951, the young biochemist Boris Pavlovitch Belooussov, who worked in the Biophysics Laboratory of the Ministry of Health of the USSR, wanted to create an inorganic reaction that was similar to the energy-producing reaction in all aerobic organisms. Such a pathway, called the "Krebs cycle" (or "citric acid cycle", abundant in lemons), works in living cells to break down sugars and produce energy. Belooussov mixed bromate ions BrO_3^- with citric acid $C_6H_8O_7$ in the presence of ceric ions Ce^{4+} in an acidic medium. He hoped to reduce cerium (4^+) to cerium (3^+), which would have caused a discoloration in the solution. However, his observations were quite different, since he noticed a periodic succession of colors and discolorations of the reaction medium: Belooussov had just discovered the first oscillating reaction by chance. Ten years later, Anatol Zhabotinsky confirmed these observations, which had previously received little credit, and in turn described these oscillating, colored and "budding" waves, which earned the two authors the Lenin Prize for the now famous *Beloousov-Zhabotinsky* reaction. These observations inspired Heinz von Foerster's (1961) and Henri Atlan's (1972) theories on self-organization, Francisco Varela's autopoiesis (1974) and Ilya Prigogine's (1947) "thermodynamics far from equilibrium", with related speculations on the origins of life.

1.3. Fossil traces?

Despite laboratory experiments and increasingly sophisticated techniques, we are faced with the absence of traces of the past, irrefutable "evidence" of the first moments. Indeed, down-to-earth, strictly geological considerations make it difficult to explore morphological evidence of the past. Volcanoes, tectonic plates, metamorphism and massive meteorite bombardments occurred during the Hadean period[1] (4.5 to 4 billion years ago), during which warm, iron-laden oceans were formed, the Archean period[2] (4 to 2.5 billion years ago), and the billions of years that preceded us.

Geological records provide direct evidence of the presence of primitive life on Earth in four main ways: through microfossils, stromatolites (literally "layers of stone", from the Greek: *stroma*, "carpet", and *lithos*, "stone"), molecular biomarkers, and stable isotope ratios. However, the interpretation of the samples and *in situ* analyses is still under discussion, as Archean rocks are scarce and poorly preserved.

Traces of ancient life may be hidden in fossil remains, which careful excavations are trying to discover. Fossils are rare because the fossilization process takes a long time, and when traces remain, they must escape decomposition and destruction, only to reappear later through soil erosion. Of the evidence left by ancient organisms, 99% is recent, dating back 545 million years. They include leaves, footprints, shells, bones, teeth, and spores or skeletons of fish or dinosaurs, most often preserved by sedimentation.

However, micropaleontologists refer to persistent traces of microorganisms in the form of stromatolites that existed almost 4 billion years ago (Schopf 1993; Allwood *et al.* 2006).

So, what happened in the first billion years of Earth's history that led to the appearance of organisms that are similar to today's bacteria?

1 Hadean: comes from the name Hades, which refers to the Greek god of the underworld.

2 Archean: from the ancient Greek *Αρχή/Arkhē*, which means "beginning, origin".

Today, microbial communities build laminated "stone mats" in two ways. The cyanobacteria's method is to trap fine sediments with a sticky film of mucus that each cell secretes, then cement the sediment grains with calcium carbonate precipitated in water. Because living cyanobacteria are photosynthetic, they move towards the light, so the bacterial mat always remains on the outside of the stromatolite. The cyanobacteria's second method of constructing stromatolites is the precipitation of their own carbonate structure in water, with the incorporation of sediments (Awramik 1994).

Corrugated forms, resulting from probable laminations, have been observed in the south-west of Greenland at sites dating back 3.8 billion years. In the Isua region, the outcrops of ancient sedimentary rocks show forms in the metamorphosed minerals that resemble layers of stromatolites 1 to 4 cm high, produced by microorganisms that would have developed at that time, as suggested by the presence of yttrium (Y), a rare earth typical of sediments deposited in a shallow marine environment. However, these results have been contested by some, who instead point to formations of tectonic origin (Nutman 2016).

Elsewhere, in the Pilbara region (Warrawoona rock formation) in Western Australia, small sedimentary cushions, or domes, aged at 3.5 billion years old can be found. Early life environments in the Pilbara Craton also included shallow marine sedimentary environments and hydrothermal regions. Other such formations, dating back 2.5 billion years, can be found in the Transvaal in South Africa.

Figure 1.1. *Stromatolites dating back 2.5 billion years, observed in the Transvaal in South Africa. Courtesy of Pierre Thomas (2016). For a color version of this figure, see www.iste.co.uk/chapouthier/life.zip*

This variety of environments, combined with the variety of stromatolite forms found, would indicate a biosphere that was already diverse 3 to 4 billion years ago.

Biodiversity originates from the origins of life itself.

A major contribution highlights a point that is rarely addressed in the study of the metabolism of early stromatolites. Indeed, the hypothesis that cyanobacteria formed fossil stromatolites by mineralization and lithification of microbial mats assumes that oxygenic photosynthesis is a very old process, one that was active more than 3 billion years ago. However, the surface of the Earth was predominantly anoxic in the Archean period, containing less than 1% of today's oxygen concentration. The increase in oxygen in the ocean, as a result of cyanobacterial photosynthesis, gradually turned it into an oxidizing environment, whereas it was initially reductive. This means that the oldest stromatolites are the product of phototrophic anoxygenic microorganisms, which do not produce oxygen. Supporting this hypothesis, researchers have recently observed that phototrophic sulpho-oxidizing alpha-proteobacteria are responsible for the precipitation of aragonite, the main calcium carbonate constituting the stromatolites of Lake Dziani Dzaha in Mayotte (Gérard *et al.* 2018).

In addition, other anaerobic methanotrophic microbial communities were present prior to the oxygenation of the Earth's atmosphere. Using sulfate ions or organic sulfur to oxidize methane, these microorganisms that lived in sedimentary lake environments left traces in the Tumbiana formation, aged at 2.7 billion years old (Lepot *et al.* 2019).

The anoxygenic photosynthesis forming the first stromatolites was carried out by phototrophic anoxygenic microorganisms and would therefore have occurred before oxygen photosynthesis.

1.4. Geochemical elements confirming these recent results

The vast majority of the world's iron is known as Banded Iron Formation (BIF). Archean banded-iron deposits are marine sedimentary rocks that are very rich in iron and today account for 90% of the iron mined in the world.

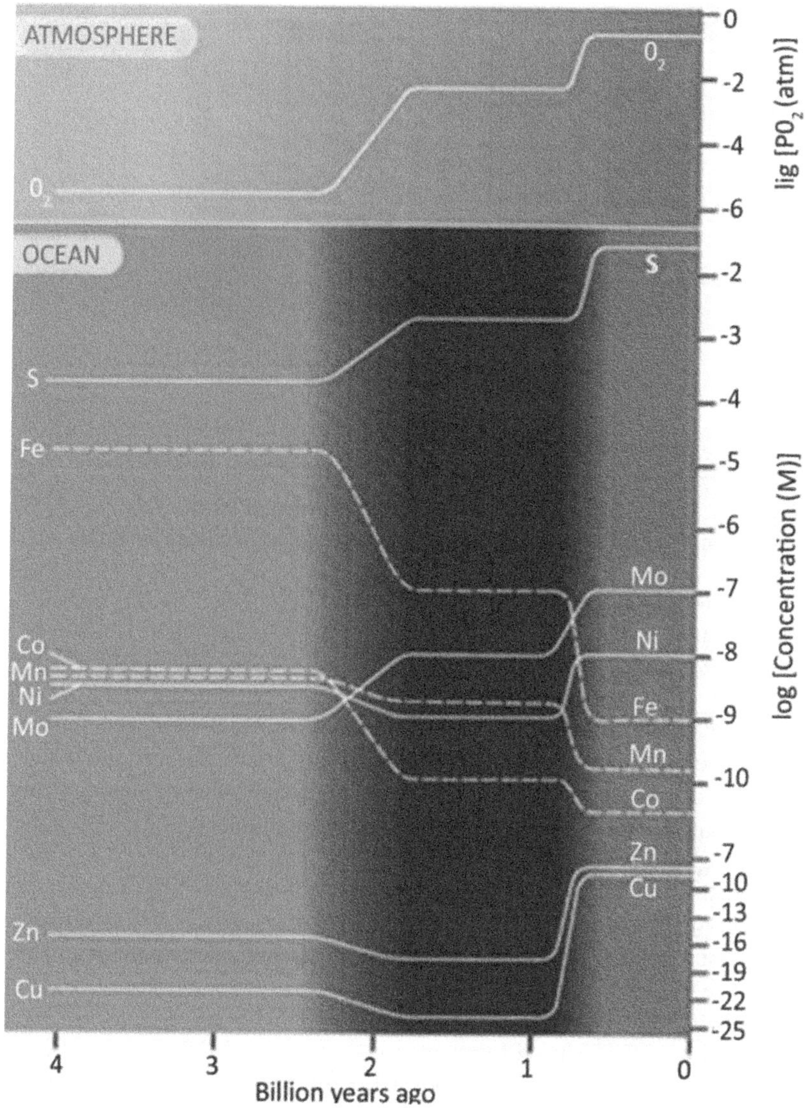

Figure 1.2. *The figure shows changes in the abundance of elements over time, mainly sulfur (S) and iron (Fe). The color gradations indicate a transition from anoxic oceans, e.g. low in sulfur, before 2.4 billion years (light blue) to oceans rich in H_2S between 1.8 billion and 800 million years (dark blue), and then to complete oxygenation of the oceans (green). Courtesy of Ariel Anbar (2008). For a color version of this figure, see www.iste.co.uk/chapouthier/life.zip*

Figure 1.3. *Banded Iron Formation (BIF). Courtesy of Pierre Thomas (2011). For a color version of this figure, see www.iste.co.uk/chapouthier/life.zip*

The ocean and the Earth's surface were without oxygen 4 to 2.5 billion years ago.

The weathering of minerals from iron-rich continents produced ferrous ions (Fe^{2+}) that were soluble in water, and therefore particularly mobile, and which were able to spread into the oceans. Volcanic activity at hydrothermal springs may also have contributed to the presence of ferrous ions in solution.

Oxygenation of the oceans by the oxygenic photosynthesis of cyanobacteria, up to about 2.4 billion years ago, caused soluble ferrous iron (Fe^{2+}) to disappear by oxidation into insoluble ferric iron (Fe^{3+}), which precipitated as magnetite and hematite.

When most of the reduced forms of iron were oxidized in the Paleoproterozoic era, sedimentation of banded iron deposits became rare. As a result, the O_2 content first increased in the oceans, then in the atmosphere, becoming toxic to anaerobic organisms. This was the Great Oxidation or "Oxygen Catastrophe".

Given that sea iron precipitated in an insoluble form (Fe^{3+}) in the Archean era, the sea water of that time contained iron in solution, in a soluble form (Fe^{2+}). This proves that the sea of that time was reduced, as was the overlying atmosphere.

Another type of photosynthesis, which is rare and little known, could explain an abundant precipitation of iron oxide: photoferrotrophy, a process where iron provides electrons.

Photoferrotrophy is a photosynthesis (less energy efficient than conventional photosynthesis) that oxidizes the iron (Fe^{2+}) of FeO into iron (Fe^{3+}) of Fe_2O_3; it can be written in a very simplified way: 2 FeO + H_2O + photons → Fe_2O_3 + 2 H^+ + 2 e^-.

The H+ ions and e^- electrons are then used by mechanisms, similar to those of classical photosynthesis, to synthesize carbohydrates from CO_2; this metabolism requires the presence of iron (Fe^{2+}) in the environment and leads to the massive precipitation of hematite Fe_2O_3.

Life is determined by the environment and, as multiple environments coexist, the origins of life and biodiversity coincide and evolve together.

1.5. Compartmentalization of resources and primary biomass

In September 1969, a fireball exploded in the sky over Murchison, Australia, followed by a shower of meteorite fragments gathered in a few days. Extraterrestrial amino acids and hydrocarbons in the Murchison meteorite were quickly identified by David Deamer (1985), who showed that organic compounds in the meteorite could also assemble into membranes. Since then, many molecules of biological interest have been identified in other meteorites (Callahan *et al.* 2011).

Figure 1.4. *Carbonaceous chondrite. Courtesy of Pierre Thomas (2016). For a color version of this figure, see www.iste.co.uk/chapouthier/life.zip*

Deamer and Barchfeld (1982), Deamer and Pashley (1989), Deamer (1997), Dworkin *et al.* (2001), will be pioneers in the experimental study of organic compounds synthesized in space. Entering into the planetary atmosphere of the primitive Earth, they mixed with endogenous species, some of them amphiphilic, with polar and non-polar groups on the same molecule. They spontaneously self-assemble to form more complex bimolecular structures which, in turn, form membrane vesicles.

The self-assembly of amphiphilic compounds, lipids forming spherical vesicles, called "liposomes", are capable of capturing macromolecules. The lipid bilayer is sufficiently permeable to allow exchanges with ionic and polar compounds from outside the compartment, allowing polymerization reactions, a kind of proto-metabolism, within these vesicles (Zepik *et al.* 2007). Hydrothermal sites are good candidates for the realization of such prebiotic evolution on the primitive Earth. Vesicles formed on mineral surfaces capture and produce diverse molecular systems. Each vesicle represents a protocell, a kind of chemical microreactor (Damer and Deamer 2015).

Laboratory simulations show that such vesicles easily encapsulate functional macromolecules, including nucleic acids and enzymes. RNA-type polymers are synthesized non-enzymatically in the laboratory from mononucleotides in lipid environments. RNA-type polymers identified by nanopore were analyzed by standard enzyme labeling methods, followed by gel electrophoresis. Chemical activation of the mononucleotides is not required. Instead, the synthesis of phosphodiester bonds is stimulated by the chemical potential of fluctuating anhydrous and hydrated conditions, with heat providing activation energy during the dehydration. In the final hydration step, the RNA-type polymer is encapsulated in lipid vesicles. This process provides the model for a possible first step in the evolution towards an RNA world (Rajamani *et al.* 2008).

Toppozini and her colleagues (2013) used X-ray diffraction to analyze mixtures of dehydrated self-assembled lipid multilamellar structures in the presence of a layer of mononucleotides such as adenosine monophosphate (AMP). The multilamellar structures are organizational structures, so condensation reactions of the mononucleotides into RNA-like polymers can thus occur (De Guzman *et al.* 2014; Misuraca *et al.* 2017).

Cell blanks were obtained by assembling phospholipids, amphiphilic molecules that interact through a hydrophilic head (which likes water) and a hydrophobic tail (which avoids any interaction with water). This structure allows organization in water in the form of vesicles. Researchers have shown that prebiotic compounds, such as polyprenyl phosphate, combine in water to form oriented double layers (Nakatani *et al.* 2014).

Jeff Errington (2013) followed the development of the *Bacillus subtilis* L-form (wall-free) bacterium and a liposome under the same conditions. Both bacteria and liposome split, fuse together and produce offspring in the form of small vesicles, which are morphologically very similar (see Figure 1.5). The unusual mode of proliferation of these bacteria by budding and fission could serve as a model for the in vitro simulation of vesicles or primitive cells. The binary fission mechanism provides insight into how early cell life forms were able to proliferate. Later in evolution, the invention of the protective cell wall may have been responsible for the expansion of bacterial life on Earth.

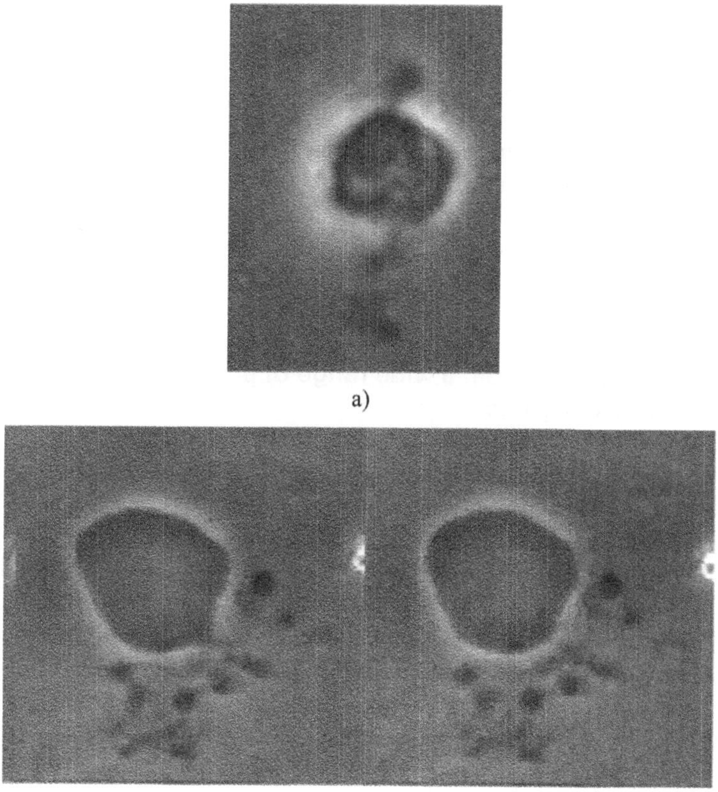

a)

b)

Figure 1.5. *a) Proliferation of L-Bacillus subtilis. Courtesy of Jeff Errington. b) Liposome budding. Phospholipid vesicles in the presence of oleic acid. Courtesy of Primož Peterlin (2009)*

Luisi and his collaborators were pioneers in the study of the processes by which lipid vesicles could develop and even reproduce (Walde *et al.* 1994a, 1994b; Luisi 2002, 2016). For example, it has been shown that when oleoyl anhydride was exposed to an aqueous buffer at alkaline pH, the anhydride was slowly hydrolyzed and the resulting oleic acid assembled into membrane vesicles. The vesicles can grow by the addition of fatty acids in the environment and then divide into smaller vesicles that grow, compete for resources, and even have a simple version of a feedback cycle that can regulate growth (Chen and Szostak 2004; Mansy and Szostak 2008; Adamala and Szostak 2013a, 2013b; Engelhart *et al.* 2016; Jimbo *et al.* 2016).

What can we deduce from these observations regarding our knowledge of the reproduction of the ancestors of today's bacteria?

Briers *et al.* (2012) studied the gap between the first protocell and the hypothetical cell ancestor named "Luca". Although there are striking coincidences between the two behaviors, the similarities observed still deserve long studies to confirm an ancestral morphological link and the understanding of the origins of unicellular life.

1.6. Rebuilding a living cell: a wide range of possibilities explored, from the mineral to the organic

Today, we speak of "chemical gardens" and "metallic vegetation" to refer to self-organized chemical figures that evoke plant diversity. When a metal salt is brought into contact with a basic aqueous solution of silicate, carbonate or phosphate, patterns called "flowers" are produced in the laboratory (Haudin *et al.* 2018). Technological innovations are suggested by these natural mineral productions which flourish today in the field of biomimicry. They find multiple biophysical, robotic, electronic, automatic, aeronautical, architectural, etc. applications, with the purpose of compensating for our weaknesses and repairing handicaps. Let us note here that mimicry is not the prerogative of modern technologies. A spider can be mistaken, even by the greatest arachnologist, for a fragment of a dead leaf.

The mechanobiology of biological membranes has become a growing field of research, particularly in the field of synthetic biology, towards artificial cells with genetic circuits and reaction cascades. Maximizing the modularity of their design and their flexibility is made possible by encapsulating them in liposomes, allowing chemical reactions to take place in well isolated environments. Such minimal synthetic cells, called “synells”, were designed by MIT researchers[3], who founded Synlife in 2017. They are governed by external signals and communication between liposomes, which can be fused in a controlled manner (Adamala *et al.* 2016).

Artificial life, in silico, a new paradigm between the machine and the living, serves as a model for the design of mineral “creatures”, organoids, automata (mechanical or bionic), micro-robots with metabolisms (Hamada *et al.* 2019), humanoids, etc., which would “emerge” from nanotechnology. The pairing of the brain and machine aims at the illusory perfection of the human species aspiring, as described by Aldous Huxley in 1932 in *Brave New World*, to impose its path to evolution.

1.7. Conclusion

Seeking to understand how life appeared on Earth also means looking for the fossil traces of the first cells, which inevitably leads to the question of how big the smallest living organism is (implied: the smallest being the simplest, therefore the first to appear on Earth). In connection with genomics work, Craig Venter (director of the Institute for Genomic Research – TIGR[4] – in Rockville, Maryland) and Hamilton Smith (who won the Nobel Prize in 1978 for the co-discovery of restriction enzymes) have removed the genes of *Mycoplasma genitalium* one by one, reputed to have the smallest known genome. The aim is to obtain a viable organism with the minimum number of genes, *Mycoplasma laboratorium*, making it possible to develop, in fine, the smallest living organism possible: “The goal is to fundamentally understand the components of the most basic living cell” (Singer 2007). This approach raises one of the most delicate problems in contemporary biology: what consequences can the manufacture of such objects have on the future of humanity, the biosphere, the environment and the planet? Is there a risk of unbalancing ecosystems?

Producing organisms that are “outside the norm”, in the technical sense, employs us to think about the standardization of criteria for living things, the quality of life, and lastly, the normativity of living things (what does “being normal” mean?). In 1995, Jean-Pierre Séris declared that it was impossible to set standards of vital value: “Uniformity, the reproduction of that which is identical has a meaning and a technical value. It has no biological value. It is biological nonsense ...”

---

3 MIT: Massachusetts Institute of Technology.

4 TIGR: The Institute for Genomic Research.

1.8. Acknowledgements

Our thanks go to Ariel Anbar, Jeff Errington, Primož Peterlin, Pierre Thomas and Peter Walde, who kindly provided us with the figures and photographs (Figures 1.1 to 1.5), and to Jacques Vergne and Louis Ter-Ovanessian for their careful reading.

1.9. References

Adamala, K., Szostak, J.W. (2013). Competition between model protocells driven by an encapsulated catalyst. *Nature Chemistry*, 5(6), 495–501.

Adamala, K.P., Martin-Alarcon, D.A., Guthrie-Honea, K.R., Boyden, E.S. (2016). Engineering genetic circuit interactions within and between synthetic minimal cells. *Nature Chemistry*, 9, 431–439.

Adriaenssens, E.M., Sullivan, M.B., Knezevic, P., van Zyl, L.J., Sarkar, B.L., Dutilh, B.E., Alfenas-Zerbini, P., Lobocka, M., Tong, Y., Brister, J.R., Moreno Switt, A.I., Klumpp, J., Karam Aziz, R., Barylski, J., Uchiyama, J., Edwards, R.E., Kropinski, A.M., Petty, N.K., Clokie, M.R.J., Kushkina, A.I., Morozova, V.V., Duffy, S., Gillis, S., Rumnieks, J., Kurtböke, I., Chanishvili, N., Goodridge, L., Wittmann, J., Lavigne, R., Bin Jang, H., Prangishvili, D., Enault, F., Turner, D., Poranen; M.M., Oksanen, H.M., Krupovic. M. (2020). Taxonomy of prokaryotic viruses: 2018–2019 update from the ICTV Bacterial and Archaeal Viruses Subcommittee. *Archives of Virology*, 165, 1253–1260.

Allwood, A.C., Walter, M.R., Kamber, B.S., Marshall, C.P., Burch, I.W. (2006). Stromatolite reef from the early Archaean era of Australia. *Nature*, 441, 714–718.

Anbar, A.D. (2008). Elements and evolution. *Science*, 322, 1481–1483.

Atlan, H. (1972). *L'Organisation biologique et la théorie de l'information*. Hermann, Paris.

Awramik, S.M. (1994). The history and significance of stromatolites. In *Early Organic Evolution: Implications for Mineral and Energy Resources*, Schidlowski, M. *et al.* (eds). Springer, Berlin, 435–449.

Bernard, C. (1878). *Leçons sur les phénomènes de la vie communs aux animaux et aux végétaux*. Librairie Baillière et fils, Paris.

Briers, Y., Walde, P., Schuppler, M., Loessner, M.J. (2012). How did bacterial ancestors reproduce? Lessons from L-form cells and giant lipid vesicles. *Bioessays*, 34, 1078–1084.

Bungenberg de Jong, H.G. (1936). *La coacervation. Les coacervats et leur importance en biologie*. Hermann & cie, Paris.

Callahan, M.P., Smith, K.E., Cleaves II, H.J., Ruzicka, J., Stern, J.C., Glavin, D.P., House, C.H., Dworkin, J.P. (2011). Carbonaceous meteorites contain a wide range of extraterrestrial nucleobases. *PNAS, USA*, 108(34), 13995–13998.

Chen, I.A., Szostak, J.W. (2004). Membrane growth can generate a transmembrane pH gradient in fatty acid vesicles. *PNAS, USA*, 101(21), 7965–7970.

Damer, B., Deamer, D. (2015). Coupled phases and combinatorial selection in fluctuating hydrothermal pools: a scenario to guide experimental approaches to the origin of cellular life. *Life (Basel)*, 5(1), 872–887.

Deamer, D.W. (1985). Boundary structures are formed by organic components of the Murchison carbonaceous chondrite. *Nature*, 317, 792–794.

Deamer, D.W. (1997). The first living systems: a bioenergetic perspective. *Microbiol. Mol. Biol. Rev.*, 61, 230–261.

Deamer, D.W., Barchfeld, G.L. (1982). Encapsulation of macromolecules by lipid vesicles under simulated prebiotic conditions. *J. Mol. Evol.*, 18, 203–206.

Deamer, D.W., Pashley, R.M. (1989). Amphiphilic components of carbonaceous meteorites. *Orig. Life Evol. Biosph.*, 19, 21–33.

De Guzman, V., Vercoutere, W., Shenasa, H., Deamer, D. (2014). Generation of oligonucleotides under hydrothermal conditions by non-enzymatic polymerization. *J. Mol. Evol.*, 78(5), 251–262.

Dworkin, J.P., Deamer, D.W., Sandford, S.A., Allamandola, L.J. (2001). Self-assembling amphiphilic molecules: synthesis in simulated interstellar/precometaryices. *Proc. Natl. Acad. Sci. USA*, 98, 815–819.

Engelhart, A.E., Adamala, K.P., Szostak, J.W. (2016). A simple physical mechanism enables homeostasis in primitive cells. *Nat. Chem.*, 8(5), 448–53.

Errington, J. (2013). L-form bacteria, cell walls and the origins of life. *Open Biol*, 3, 120–143.

Fox, S.W. (1988). *The Emergence of Life*. Basic Books Publishers, New York.

Gérard, E., De Goeyse, S., Hugoni, M., Agogué, H., Richard, L., Milesi, V., Guyot, F., Lecourt, L., Borensztajn, S., Joseph, M.B., Leclerc, T., Sarazin, G., Jézéquel, D., Leboulanger, C., Ader, M.K. (2018). Role of alphaproteobacteria and cyanobacteria in the formation of stromatolites of Lake Dziani Dzaha (Mayotte, Western Indian Ocean). *Frontiers in Microbiology*, 9.

Haldane, J.B.S. (1929). The origin of life. *The Rationalist Annual*, 3–10.

Hamada, S., Yancey, K.G., Pardo, Y., Gan, M., Vanatta, M., An, D., Hu, Y., Derrien, T.L., Ruiz, R., Liu, P., Sabin, J., Luo, D. (2019). Dynamic DNA material with emergent locomotion behavior powered by artificial metabolism. *Science Robotics*, 4(29).

Haudin, F., Brau, F., De Wit, A. (2018). La chimie génératrice de forme : végétation métallique et jardins chimiques [Online]. *Revue Arts et Sciences V2 N1*. Available at: https://www.openscience.fr/Numero.

Jimbo, T., Sakuma, Y., Urakami, N., Ziherl, P., Imai, M. (2016). Role of inverse-cone-shape lipids in temperature-controlled self-reproduction of binary vesicles. *Biophysical Journal*, 110, 1551–1562.

Leduc, S. (1912). *La biologie synthétique*. Poinat, Paris.

Lepot, K., Williford, K.H., Philippot, P., Thomazo, C., Ushikubo, T., Kitajima, K., Mostefaoui, S., Valley, J.W. (2019). Extreme ^{13}C depletions and organic sulfur content argue for S-fueled anaerobic methane oxidation in 2.72 Ga old stromatolites. *Geochimica et Cosmochimica Acta*, 244, 522–547.

Locey, K.J., Lennon, J.T. (2016). Scaling laws predict global microbial diversity. *PNAS*, 113(21), 5970–5975.

Luisi, P.L. (2002). Toward the engineering of minimal living cells. *The Anatomical Record*, 268, 208–214.

Luisi, P.L. (2016). *The Emergence of Life*. Cambridge University Press, Cambridge.

Mansy, S.S., Szostak, J.W. (2008). Thermostability of model protocell membranes. *PNAS, USA*, 105(36), 13351–13355.

Misuraca, L., Natali, F., Da Silva, L., Peters, J., Demé, B., Ollivier, J., Seydel, T., Laux-Lesourd, V., Haertlein, M., Zaccai, G., Deamer, D., Maurel, M.C. (2017). Mobility of a mononucleotide within a lipid matrix: A neutron scattering study. *Life (Basel)*, 7(1).

Mora, C., Tittensor, D.P., Adl, S., Simpson, A.G.B., Worm, B. (2011). How many species are there on Earth and in the ocean? *PLoS Biol.*, 9(8).

Nakatani, Y., Ribeiro, N., Streiff, S., Gotoh, M., Pozzi, G., Désaubry, L., Milon, A. (2014). Search for the most 'primitive' membranes and their reinforcers: a review of the polyprenyl phosphates theory. *Orig. Life Evol. Biosph.*, 44(3), 197–208.

Nutman, A.P., Bennett, V.C., Friend, C.R.L., Van Kranendonk, M.J., Chivas, A.R. (2016). *Nature*, 537, 535–538.

Oparin, A.I. (1938). *The Origin of Life*. MacMillan, New York.

Peterlin, P., Arrigler, V., Kogej, K., Svetina, S., Walde, P. (2009). Growth and shape transformations of giant phospholipid vesicles upon interaction with an aqueous oleic acid suspension. *Chem. Phys. Lipids*, 159(2), 67–76.

Prigogine, I. (1947). *Étude thermodynamique des phénomènes irréversibles*. Dunod, Paris.

Rajamani, S., Vlassov, A., Benner, S., Coombs, A., Olasagasti, F., Deamer, D. (2008). Lipid-assisted synthesis of RNA-like polymers from mononucleotides. *Orig. Life Evol. Biosph.*, 38, 57–74.

Schopf, J.W. (1993). Microfossils of the Early Archean Apex chert: New evidence of the antiquity of life. *Science*, 260(5108), 640–646.

Séris, J.-P. (1994). *La technique*. PUF, Paris.

Singer, E. (2007). Craig Venter's genome. *MIT Technology Review*, September 4.

Thomas, P. (2011). Les fers rubanés (*Banded Iron Formation*) de l'Archéen de Barberton, groupe de *Fig Tree* (-3,26 à -3,22 Ga), Afrique du Sud [Online]. Available at: https://planet-terre.ens-lyon.fr/image-de-la-semaine/ Img364-2011-10-10.xml.

Thomas, P. (2016). L'origine de la vie vue par un géologue qui aime l'astronomie [Online]. Available at: https://planet-terre.ens-lyon.fr/article/origine-vie-2016.xml.

Toppozini, L., Dies, H., Deamer, D.W., Rheinstädter, M.C. (2013). Adenosine monophosphate forms ordered arrays in multilamellar lipid matrices: Insights into assembly of nucleic acid for primitive life. *PLoS ONE*, 8(5).

Varela, F.G., Maturana, F.H., Uribe, R. (1974). Autopoiesis: The organization of living systems, its characterization and a model. *BioSystems*, 5(4), 187–196.

Von Foerster, H. (1961). A predictive model for self-organizing systems. *Cybernetica*, 3(4), 258–300.

Walde, P., Goto, A., Monnard, P.-A., Wessicken, M., Luisi, P.L. (1994a). Oparin's reactions revisited: enzymatic synthesis of poly(adenylic acid) in micelles and self-reproducing vesicles. *J. Am. Chem. Soc.*, 116, 7541–7547.

Walde, P., Wick, R., Fresta, M., Mangone, A., Luisi, P.L. (1994b). Autopoietic self-reproduction of fatty acid vesicles. *J. Am. Chem. Soc.*, 116, 11649–11654.

Zepik, H.H., Rajamani, S., Maurel, M.C., Deamer, D. (2007). Oligomerization of thioglutamic acid: encapsulated reactions and lipid catalysis. *Orig. Life Evol. Biosph.*, 37, 495–505.

2

The Nature of Life

Andreas LOSCH
Faculty of Theology, University of Bern, Switzerland

How can we characterize the nature of life? Against other assumptions, science has shown that life is nothing but physics and chemistry, and that there is no mysterious vital force included. Still, life strikes us as special. Can we describe this by its form? What definitions of life are actually in use? Maybe we need another sort of language to describe life's specialty.

This chapter will deal with the question of the nature of Life on Earth, so life-as-we-know it. The astrobiological quest for the search for extraterrestrial life is certainly very worthwhile to pursue, as in the case of success we would have a greater sample of life than only the one sharing common ancestry on Earth and hence would probably learn a great deal more about the laws of nature governing the emergence of life (Sagan 2010, p. 306). Does every life need to follow the "right-handed" chirality (orientation) of its molecules employed by Earth life, for instance? Does life need to be carbon-based or is this only valid for its Earthly form?

In the current absence of such further samples, I believe it, however, to be most important to stick to what we know about Earth life to attempt at understanding more of its very nature. We do not know so very much about its origins and its essence, besides that it is physics and chemistry, with an evolutionary history, for sure (Lazcano 2017b, p. 90).

The Explosion of Life Forms,
coordinated by Georges CHAPOUTHIER and Marie-Christine MAUREL. © ISTE Ltd 2020.

2.1. Observations and assumptions

Maybe asking about life's "essence" is, however, already the wrong question. In traditional philosophical terms, asking about the essence of something aims at finding out its "formal cause", a concept not in use in modern science. It only eventually plays a role when it comes to the idea of "information". For Aristotle's philosophy, every life even has some sort of *soul*, which also involves another, *final cause* or *purpose* for an object in the world (Bedau and Cleland 2010, p. 4). Although this is distinct from the widespread bodiless concept of soul advocated by René Descartes, the whole idea behind it does not fit into modern science anymore. When we ask about the nature of something, like in this very short essay, we want to know of the substance of something, not its purpose, which for Aristotle, in contrast, was also part of its nature. Physics and metaphysics were combined.

The idea of teleological or goal directed attributes of life has reappeared from time to time in the discussion. Immanuel Kant, for instance, was a very prominent proponent of the idea of teleology. Even Darwin seems to have made some use of his version of the concept, but only in his notebooks (Lennox 1993). More recently, but still before the dawn of mature Darwinism, embryologist Hans Driesch turned into a philosopher when postulating a vital force as being responsible for the possibility of two perfect embryos developing from an egg separated into two after the first cleavage division (Mayr 1982, p. 118).

The approach has been discarded by a vast majority of biologists, who are keen to point out that there is no entelechy at work which distinguishes life from non-life. Ernst Mayr, for instance, would acknowledge some sort of teleonomy in individuals, namely systems operating based on a program of coded information (Mayr 1965, p. 42), but "there is no evidence that would support teleology in the sense of various vitalistic or finalistic theories" (Mayr 1965, p. 43). I suppose those theories might be projections from how we ourselves perceive the world as assumingly willful beings.

Although the Aristotelian world view has been an empirical approach, its conclusions were extrapolations from how things appeared to us, from our own observations. Reality can be trickier than that and we sometimes must leave our own point of view: to understand that the sun does not revolve around Earth, for instance, although observationally it may at first look like it when it rises and sets. Only more careful observations of the celestial constellations made the emergence of the heliocentric world view possible, even before we could leave our Earthbound perspective behind.

Our culture is also influenced by its religious context. The theological idea of a Divine breath of life animating bodies (Gen 2:7; Ps 104:29f) and granting the ability to live may be in part another observation-based approach which fails. We observe that when a human's or animal's breathing ends, it is dead, so the invisible breath seems to constitute the difference. An apparently invisible force is associated here with the Divine.

This may remind us that our language carries with it a certain world-view and tradition and we must be cautious about the words we use and sometimes question their traditional annotations. We know today that the observation of all life performing breathing or even just requiring air in a strict sense is certainly wrong, as life must have already emerged in an environment when there was no oxygen-rich atmosphere on Earth (Catling 2013, p. 32). Yet some ideas of religious thinkers might also prove helpful to develop science. The whole idea of evolution, for instance, is associated with the idea of a linear development and progress typical for the Judeo-Christian tradition, for instance (Lazcano 2017a, p. 311). Likewise, *metabolism* as a more general approximation to the concept of the breath of life is probably a justified observation to stick to.

2.2. Descriptions and definitions

What is life? Do we really know? The way to define life as long as we did not know how it functions was to describe the conglomerate of attributes which characterize it. A similar approach led to describing water according to its properties – like "being wet, transparent, odorless, tasteless, thirst quenching, and a good solvent", which was a helpful approximation until we found out its H_2O structure, which is the scientifically most informative answer to define water today (Cleland and Chyba 2010, p. 326).

Similar, a (1) *physiological* approach to define life, long-time popular, describing life according to all of its functions – like eating, metabolizing, excreting, breathing, moving, growing, reproducing, and being responsive to external stimuli – was a first approximation, but is less in use today, as "such properties are either present in machines that nobody is willing to call alive, or absent from organisms that everybody is willing to call alive" (Sagan 2010, p. 303). Automobiles share too many of these properties of the "living" and are certainly not alive. Also, some obviously living bacteria do not breathe at all.

More promising appears to be an approach focused on (2) *metabolism*, describing "a living system as an object with a definite boundary, continually exchanging some of its materials with its surroundings, but without altering its general properties, at least over some period of time" (Sagan 2010, p. 303). Here as well, however, some known examples of life do not fit into the picture. What about seeds, which can lay dormant quite a while before awakening to revive? A very popular counterexample is also fire, which is a metabolism that is, in addition, even showing growth. Clearly, metabolism alone – while a *necessary* property of all actual life – cannot be *sufficient* to describe its nature.

Maybe a (3) *biochemical* description does better. Physicist Erwin Schrödinger had speculated about an "*aperiodic crystal*" being an essential ingredient to chromosomes and hence to life (Schrödinger 1992, p. 5). Even more accurate stipulations regarding molecular information storage have been made earlier by Nikolai Koltsov (Soyfer 2001, p. 726). As Watson and Crick did not notice Koltsov, it was Schrödinger's speculations which made Crick think "that great things were just around the corner" (Crick 1990, p. 18), and as is well-known, Crick's efforts have been crowned by the discovery of the double-helix structure of the DNA. So, something like Schrödinger's *aperiodic crystal* did exist. Today, "reproducible hereditary information coded in nucleic acid molecules" (Sagan 2010, p. 304) is seen as a key ingredient of life, but still, there are exceptions even to this characterization. What about viruses without a DNA of their own? (On this question see Morange 2017, 103f).

A (4) *genetic* definition of life would be "a system capable of Darwinian evolution by natural selection"; it takes into account the progressive development of life forms. Even without the assumption of an increasing complexity throughout evolutionary history, the evolutionary scenario is appealing. How so? It simply seems to make sense: in the moment self-replicating units emerged on Earth, those units filled their respective environment. Any replication carries with it the possibility for duplication errors, which leads (besides a lot of "waste") to mutant versions of the original unit. Those mutants replicating more efficiently replicated more quickly. Limited resources provided, this leads to the dominance of particular replicators over others in a given environment, while some mutants were able to make use of new environments. As a result, Earth is soaked through with life, even in extreme environments totally hostile to human life, microbes (called extremophiles) exist.

While the genetic definition seems to stay valid and is in use today, there is a further way of defining life via (5) *thermodynamics*. Also, Schrödinger had hinted towards something like negative entropy as life's property (Schrödinger 1992,

p. 70). What does this mean? Normally, according to the laws of thermodynamics, no processes can occur that increase the net order of a given system. Structures tend to dissolve over time, entropy is increasing, and our universe develops into a state of disorder, called the *heat death* of the universe (which is an extremely *cold* state because matter is so spread out). How, then, can the scale of order show us which life forms it is? The reason for this is that what has been said is only valid for closed systems. Living systems are, in contrast, *open* systems and create localized regions with an increasing order. The difference between open and closed systems is the open system's exchange with their environment. As we looked at *metabolism* as a characteristic even to define it, life certainly has this property and is an open system. Think of sunlight nourishing the plants, for instance. Now, this local increase in order comes at the cost of an overall increase in disorder (entropy). The sun loses energy steadily by fueling the thermonuclear processes necessary to radiate and shine on Earth. This way, the laws of thermodynamics are still valid, while local order is allowed to grow. Life could be described as an open system. There is a counterexample to this thermodynamic definition of life, however, crystals, which are open systems as well, are not considered living entities (Cleland and Chyba 2010, p. 327).

So, what is life? Popular today is a version of the genetic or Darwinian definition, characterizing life as a "self-sustained chemical system capable of undergoing Darwinian evolution" (Joyce *et al.* 1994, xi), including the processes of "self-reproduction, material continuity over a historical lineage, genetic variation, and natural selection" (Cleland and Chyba 2010, p. 328). This definition excludes any potential life based on other components than chemistry-based Earth life. This is why the quest to define life has been revived not only in the context of synthetic biology but also in that of astrobiology. The definition also excludes viruses, as they are not self-sustained but need a host to replicate.

2.3. Exploration

There are some objections to the Darwinian definition, as life could potentially be based on other components than chemistry. There could be computer or robotic life, as Dawkins (2010) and Dennett (1996) point out. Portraying life in a mechanical way certainly has some attractivity to it in the light of the modern scientific world view. If life can be described mechanically and in a purely functional way, why could it not be based on mechanics as well? As a matter of fact, however, although we do know quite a wealth of life forms on Earth, we do not know any life based on non-organic components. So, it makes sense to me to agree with Oparin that "[l]ife is a property of any organism, from the highest to the lowest, but it does not exist in inorganic natural objects, no matter how complicated their structure" (Oparin 2010, p. 71). There might be exceptions hidden in the depths of the universe, but Oparin would define them differently than "life". That being said, I believe him to be right insofar that we need to focus on Earth life to understand the phenomenon better, though the result could be transferable to other environments as well.

Much about science is about the fuzziness and continuous transition in nature apart from human taxonomies. Maybe "life" is a categorial description, which is hard to apply to aspects of nature which nevertheless play an important role in the emergence of the phenomenon. When "the origin and the evolution of life coincide" (Maurel 2017, p. 22) and life is physics and chemistry with a history, the small size and gradualness of historical transitions may explain the difficulty to "capture" the moment when life on Earth emerged. "If the origin of life is seen as the evolutionary transition between the non-living and the living, then it is also necessary to recognize it as a historical process shaped in part by the vagaries of contingent events. It also suggests that attempts to draw a strict line between these two worlds may be meaningless. In other words, the appearance of life on Earth should be seen as a non-progressive evolutionary continuum that seamlessly joins the prebiotic synthesis and accumulation of organic molecules in the primitive environment, with the emergence of self-sustaining, replicative chemical systems capable of undergoing Darwinian evolution (Lazcano 2017b, p. 90).

There still might have been accelerated transitions like later in the history of life (Eldredge and Gould 1972). We should avoid, however, demanding the emergence of life to be somewhat discontinuous in case we are only motivated by the desire to ascribe meaning and distinctiveness to our own status as living beings; we could be biased. Yet even if there is nothing special about organisms, this "does not mean an absence of significant differences deserving more thorough study by scientists" (Morange 2017, p. 107).

We have written about life as a natural phenomenon. We, the ones reading and writing this, are part of the picture. We are the ones defining and applying categories on nature by use of language with its subject-predicate-object structure. One could even ask "whether life can at all be described as 'something' common to all living beings making them living beings, whether the term 'life' realistically refers to an aspect of reality, or whether the term 'life' is only meaningful within language" (Rehmann-Sutter 2016, p. 270). Our point of view is dominated by the perspective of intelligent, conscious and speaking life for sure. Contrary to a purely constructivist approach, I would claim, however, being alive is certainly an aspect of reality as well. That it maybe cannot exactly defined or delimited does not mean it does not exist, in all the fuzziness of nature. “The recognition that life is the outcome of an evolutionary process constrained by the laws of physics and chemistry can lead to the acceptance that many properties associated with living systems, such as replication, self-assemblage or catalyses, are also found in non-living entities. Some systems may not be ‘half-alive’ but they can exhibit some of the properties we associate with living entities” (Lazcano 2018, p. 11).

2.4. Conclusion

A definition of life by a set of properties is the best we currently have. We used the example of water, to make clear that this may present a preliminary status only to describe life. Maybe we have to extend our informational perspective on life beyond genetic informationalism to understand the full phenomenon. This challenge may be more complex than initially imagined. Consider this: H_2O may be water, but not that what we mean when we ask for water. Pure H_2O – distilled water – would actually kill us; there is a lot more in the actual mix of “water” which makes it usable for us. Likewise, even if we identify the special relation of components which distinguishes living phenomena even before the emergence of RNA and DNA, we might think of something different when using “life” in our everyday language.

Maybe several difficulties would be resolved if we invent a new language specifically for living phenomena (an idea of A. Lazcano, private conversation) – like physics has mathematics as a language of its own. Even then, the complex phenomenon we as humans refer to by attributing something to be “alive” may be something else than life in purely scientific terms or language.

So, what is the nature of life? As long as we lack the appropriate words to answer it concisely, we must continue to use a conglomerate of words to describe what we mean by “life”. Organisms have a *combination* of characteristics that are never found together in inanimate objects (Morange 2017, p. 107). In these regards all the definitions listed under section (2) still have their moment of truth; to some degree, we still are at the stage of a physiological approach. On the other hand, we already know metabolism is a necessary ingredient of any life as an open system. And finally, and most important, an informational or biochemical pattern (like RNA or DNA) is of the essence to allow replication and genetic development with it.

2.5. References

Bedau, M., Cleland, C. (eds) (2010). *The nature of life: Classical and contemporary perspectives from philosophy and science* [Online]. Cambridge University Press, Cambridge. Available at: http://www.loc.gov/catdir/enhancements/fy1006/2010004399-d.html.

Catling, D. (2013). *Astrobiology: A Very Short Introduction*. Oxford University Press, Oxford.

Cleland, C., Chyba, C. (2010). Does "life" have a definition?. In *The Nature of Life: Classical and Contemporary Perspectives from Philosophy and Science*, Bedau, M., Cleland, C. (eds). Cambridge University Press, Cambridge, 326–339.

Crick, F. (1990). *What Mad Pursuit: A Personal View of Scientific Discovery*. Penguin Books, London.

Dawkins, R. (2010). Universal Darwinism. In *The Nature of Life: Classical and Contemporary Perspectives from Philosophy and Science*, Bedau, M., Cleland, C. (eds). Cambridge University Press, Cambridge, 360–373.

Dennett, D.C. (1996). *Darwin's Dangerous Idea: Evolution and the Meanings of Life*. Touchstone, New York.

Eldredge, N., Gould, S.J. (1972). Punctuated equilibria: an alternative to phyletic gradualism. In *Models in Paleobiology*, Dodd, J.R., Schopf, T.J.M. (eds). Freeman, San Francisco, 82–115.

Joyce, G.F., Deamer, D.W., Fleischaker, G.R. (eds) (1994). *Origins of Life. The Central Concepts*. Jones and Bartlett Publishers, Boston.

Lazcano, A. (2017a). A Skeptical Afterword. In *What Is Life? On Earth and Beyond*, Losch, A. (ed.). Cambridge University Press, Cambridge, 311–313.

Lazcano, A. (2017b). Precellular Evolution and the Origin of Life: Some Notes on Reductionism, Complexity and Historical Contingency. In *What Is Life? On Earth and Beyond*, Losch, A. (ed.). Cambridge University Press, Cambridge, 75–93.

Lazcano, A. (2017). The Emergence of Life: Some Notes on the Origin of Biological Information. In *Life Sciences, Information Sciences*, Gaudin, T. *et al.* (eds). ISTE Ltd, London, and Wiley, New York, 1–15.

Lennox, J.G. (1993). Darwin was a teleologist. *Biology and Philosophy*, 8, 409–421.

Maurel, M.C. (2017). Reflections on Origins, Life and the Origins of Life. In *What Is Life? On Earth and Beyond*, Losch, A. (ed.). Cambridge University Press, Cambridge, 13–29.

Mayr, E. (1965). Cause and Effect in Biology. In *Cause and Effect. The Hayden Colloquium on Scientific Method and Concept*, Lerner, D. (ed.). Free Press, New York, 33–50.

Mayr, E. (1982). *The Growth of Biological Thought: Diversity, Evolution, and Inheritance*. Belknap Press, Cambridge.

Morange, M. (2017). Science and Philosophy Faced with the Question of Life in the Twenty-first Century. In *What Is Life? On Earth and Beyond*, Losch, A. (ed.). Cambridge University Press, Cambridge, 97–110.

Oparin, A. (2010). The Nature of Life. In *The Nature of Life: Classical and Contemporary Perspectives from Philosophy and Science*, Bedau, M., Cleland, C. (eds). Cambridge University Press, Cambridge, 70–87.

Rehmann-Sutter, C. (2016). Welches Leben erfasst die Molekularbiologie? In *Das Leben: Historisch-systematische Studien zur Geschichte eines Begriffs*, Schaede, S., Anselm, R., Köchy, K. (eds). Mohr Siebeck, Tübingen, 261–281.

Sagan, C. (2010). Definitions of Life. In *The Nature of Life: Classical and Contemporary Perspectives from Philosophy and Science*, Bedau, M., Cleland, C. (eds). Cambridge University Press, Cambridge, 303–306.

Schrödinger, E. (1992). *What Is Life? The Physical Aspect of the Living Cell: With Mind and Matter and Autobiographical Sketches*. Cambridge University Press, Cambridge.

Soyfer, V.N. (2001). The consequences of political dictatorship for Russian Science. *Nature Reviews, Genetics*, 2(9), 723–729.

3

From Form to Function

Jean-Pierre GASC
National Museum of Natural History, Paris, France

"Not only does every activity allow itself to be discerned and defined to the extent that it takes form, where it inscribes its curve in space and time; life itself acts essentially as a creator of forms. Life is form, and form is the way of life."

(Focillon 1996)

3.1. Form: a concept for knowledge

Does the term *form* refer to a concept or a phenomenon of material reality? The highlighted quotation comes from a philosopher of aesthetics, and it's worth recalling that Kant (1790) dealt in parallel with reflexive, empirical, and aesthetic judgement to propose that the notion of form finds its source in our perceptual capacity, from which we branch off towards aesthetic pleasure or towards an empirical analysis, directed towards knowledge. But is there a "pure" form, without content, without evoking anything other than a contour, such as the limit in space between two fields, a figure belonging to the geometry of early agricultural surveyors and which has become, by abstraction, a Platonic idea? In contrast, doesn't every form instead imply a more or less simple story about its origin and meaning? A temporal aspect that the figures of geometry do not present. In his treatise *Parts of Animals*, Aristotle gives form a pre-eminent place: "nature, according to form, is more important than material nature." The constituent materials are thus relegated to the background; what counts is the form of the living being. For the Greek philosopher, however, form is a configuration, that is to say, a system of relations between figures whose mainspring is their function. This

The Explosion of Life Forms,
coordinated by Georges CHAPOUTHIER and Marie-Christine MAUREL. © ISTE Ltd 2020.

interpretation will be taken up as a fundamental principle of the organization of living beings by Georges Cuvier, founder of comparative anatomy, when he states his principle of organic correlations (Gasc 2013). As for the question of the origin of forms, it is once again Kant who goes a step further by designating living beings as both cause and effect of themselves, and their parts "reciprocally cause and effect their form". In other words, they are capable of reproduction, regeneration and self-assembly, unlike non-living objects. Thus, the concept of form has been fundamental as a means of knowledge and comparison of living beings, as it was for René-Just Haüy in the classification of mineral crystallizations in 1783[1]. The search for order is a constant in the human mind. Successive methods of classification have been based on the analysis of forms in order to identify groupings by resemblance, to arrive at invariants and to anchor sets of forms around definitions, because there is no classification without the intervention of language in the form of a hierarchical nomenclature (Foucault 1966). However, in this typological quest that animated minds in the 18th and 19th Centuries, inevitable debates about the nature of reality had to be overcome: is there such a thing as a "natural" classification? The discovery of invariants with the comparative anatomical studies of Daubenton and Vicq d'Azyr in the 18th Century gave rise to the idea that such diverse forms could be nothing more than variations on one or more ideal models (Gasc 2006). Between the required discontinuity by nomenclature and the continuity suggested by the existence of invariants and possible transformations, in other words, the passage from one form to another, there was a possible mediation: the justification of forms by their functions, returning to Aristotle. This made it possible to explain the diversity of forms by a diversity of circumstances. A new aspect about time and the history of the Earth, which was already mentioned by Buffon, was then introduced. Lamarck's transformational conception, which he presented in his lectures at the museum as early as 1802, brought a theoretical and materialistic conclusion to the debates of the previous century and opened the field to biology as a historical science (Gayon 2005). The theory Charles Darwin proposed in 1859 is quite different. He called it "descent with modification". If it retains Lamarck's hereditary transmission of acquired traits, it gives the medium the role of sorting out the material offered by the spontaneous and random variation of traits, without any direction towards progress, but simply leading to the temporary success that marks the reproduction of a greater number of descendants: this is natural selection. The invariants and similarities in composition that had led the German philosophers of

1 Haüy established the first rational classification of crystalline structures on the assumption that the geometrical form reflected an elementary organization ("integrating molecules"). He escaped the massacre of the refractory priests in 1792, thanks to the intervention of Étienne Geoffroy Saint-Hilaire, who was his pupil and was to become professor of zoology at the Museum created in 1793.

nature and Etienne Geoffroy Saint-Hilaire to defend the idea of unity of composition are, for Darwin, just proof of a community of ancestry. Morphology, which according to Goethe was to be the search for ideal forms, becomes the analysis of living forms in their adult state and in their development, in order to reconstruct their history, in other words, their genealogy. This is the great period of comparative anatomy and embryology. It was dominated by Ernst Haeckel (1866), a great defender of the theory of evolution, which was paradoxically based on the founding principles of Cuvier and Von Baer, both of whom were opposed to the idea of an evolution of the living world. The first phylogenetic trees then appeared, intended to represent the entire history of living lineages. However, when Darwin died, little was known about the origins of variation and hereditary transmission. It was the role of genetics, from the early years of the 20th Century, which was able to replace the notion of "traits" with that of "factors", soon to be called "genes", whose ability to "mutate" was discovered. From then on, it was experimentation on laboratory models that gradually opened the way to understanding the laws responsible for the diversity of living forms, by not only entering into the mechanisms of variation but also into those of selection in natural conditions of existence. The inventory of forms continues and their arrangement within the framework of their kinship is gradually enriched with analyses carried out at the different levels of integration, from that of the organism to the molecule.

3.2. Basic structural elements: from the molecule to the cell

Today, the evolutionary sciences are progressing through the cooperation of a large number of disciplines, some of which are traditionally far removed from biology, such as cosmology or computer science. Living beings carry out all their activities – metabolism, growth and reproduction – in a field of forces (radiation, gravity, surface forces), and in the aspects of space and time. Archaea and bacteria, two sets of prokaryotes that shared the original aquatic environment, have elements that subject them primarily to the forces of viscosity and surface tension over their entire surface. However, by association, these microscopic beings have created a difference in scale, the formation of a volume and a mass, on which gravity and current will be exerted. More than a billion years had to pass in order to see the appearance of co-partitioned cells, the eukaryotes, with a nucleus containing DNA isolated from the cytoplasm, and sophisticated forms of unicellular beings, Radiolarians, Foraminifera, then Diatoms, whose mineralized envelopes with astonishing forms have composed sedimentary rocks by settling on the ocean floor. A new type of living being then appeared. They're built by the self-assembly of cells resulting from successive divisions, the mitoses. Thus, in some protists (Buss 1987), the cells aggregate, forming a colony within which a group of cells differentiate and give a new clone, something that is found in true multicellular organisms such as sponges (Porifora). Multicellular organisms have been formed on this principle, and some forms of these, the placozoa, remain in nature today (see Figure 3.1).

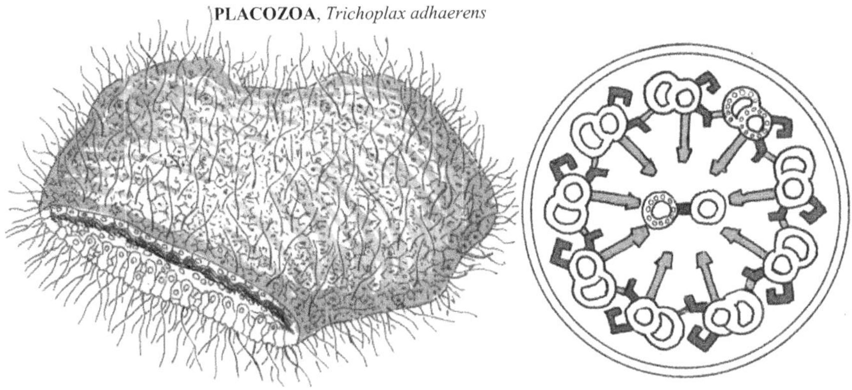

Figure 3.1. *Contractile proteins (in red) flex the microtubules (9+2) in the cilia and flagella, ensuring cell mobility of eukaryotes. For a color version of this figure, see www.iste.co.uk/chapouthier/life.zip*

Then, the union of two cells from a particular division involving the genome – the meiosis that produces gametes – inaugurates sexual reproduction. This is no longer identical multiplication, cloning, but the mixing of two individual histories. In 1885, August Weismann introduced the idea that organisms are composed of a germ cell line passing from one generation to the next and a somatic cell line providing conservation. However, this general pattern shows considerable variation among the various phylums, revealing a bushy evolution. Leo Buss gave a summary table of known cases in 1987. On the one hand, there are nematodes (roundworms), germ cells that are distinct from the beginning of ontogenesis, most often under the control of genetic agents of maternal origin, and on the other hand, there are plants, in which any somatic cell is capable of reconstituting duality by producing gametes. It has been known since 2006 that the inhibition of four genes makes it possible to reprogram differentiated cells, such as mouse fibroblasts, which return to a pluripotent state capable of producing germ cells. Opisthokonts, which include fungi (mycota) and animals (metazoans), share, at least at one stage of their lives, the characteristic of possessing a flagellum, a complex organ derived from the cilia (see Figure 3.1) located at the back of the cell, whose mobility provides the propulsion necessary for sexual reproduction. The Metazoan embryo is, first of all, a hollow ball called "blastula". The cells enter the second or third phase of a tissue differentiation, which separates the Metazoa into the first two subsets. In diploblasts, a leaflet is in direct contact with the medium, the ectoderm or ectoblast, and an inner leaflet, the endoderm, is closed in on itself by the proliferation of ectodermal cells. The continuation of these divisions leads to the gastrula stage in several ways. The endoderm is attached to the inner face of the ectoderm, delimiting a primordial intestine, the archenteron. An orifice, the blastopore, remains. Between it and the point spatially opposite it, there is a radial axis of symmetry that passes. This initial symmetry is lost in adult sponges that are tossed around by currents. In contrast, Cnidarians and Ctenophores orient their position with respect to gravitational verticality. Some of the Cnidarians settle on the bottom (benthic) after a larval phase of free life (pelagic), with the food opening lined with a crown of tentacles and the oral pole pointed upwards. They are polyps and sea anemones. At this stage, a balancing organ, the statocyst, appears, using the inertia of a mineralized corpuscle in the excitation of cilia connected to a nerve ending, transducing a mechanical effect into nerve impulses. There are no organs, but categories of specialized cells, some contractile, others mucus-producing. There are nerve cells and, in the Cnidaria, there are stinging cells and cnidocysts, each constituting a kind of mini-trap projecting a venomous harpoon. The tentacles that form a crown around the single opening in Cnidarians reflect a radial symmetry of order 4 or 6, to which is added a distribution in two groups, divided by a plane (biradial symmetry). The development goes through a mobile larval phase which then takes the form of a small jellyfish. This form of multicellular organism is known from the fossil fauna of Ediacara, dating back 680 million years. The species that form colonies of polyps form the basis of coral reefs by making an exoskeleton of calcium carbonate. The ribbons of Ctenophores are more discrete, completely transparent; they do not make a skeleton and have neurons (Bucher and Anderson 2015). They are also carnivores present in the plankton and they move with the help of eight rows of cilia that form combs arranged in a bilateral symmetry. This mode of movement using eyelashes reaches the limits of efficiency given their size. Their ribbon-like or globular form aroused the aesthetic admiration of Haeckel (1886). In triploblastic organisms, a third layer of cells is interposed between the ectoderm and endoderm, the mesoderm. Organs corresponding to a division of physiological work appear by the grouping of specialized cells into topographically determined islets. The segregation of the cells of the embryo into sheets is accompanied by the loss of radial symmetry in favor of bilateral symmetry: there is a left and a right side, as well as a dorsal and a ventral face. The cells register their position via contact glycoproteins. This general character suggests a common origin for all metazoans. This hypothesis, put forward at the time by Darwin, has been confirmed by the most recent molecular data. However, as pointed out by Buss (1987), the conservation of early stages of development cannot be explained by the selection of individuals, but depends on the effects of selection interactions operating at the various levels of biological organization, from the molecule to the organism in action in its environment.

Advances in the understanding of how genes work have benefited from molecular biology work done on prokaryotes, bacteria and viruses, which do not show compartmentalization between a nucleus and a cytoplasm. This relative simplicity has made it possible to break free from the dualistic conception (nucleus, the seat of heredity, and cytoplasm, the seat of metabolism) and to highlight the predominant role of interaction processes, a world of exchanges and the transfer of substances that become active when they meet their targets. DNA had been imagined as the citadel isolated in the nucleus, delivering its orders to the rest of the cell. In reality, not only are its messages interpreted by agents that themselves become messages, but some have feedback effects on gene expression. Ribonucleic acids (RNA) (messenger, transfer and ribosomal) are the main agents, circulating through the cell's cytoskeleton, its microtubules and the properties of the molecular motors (the contractile molecules of actin and myosin). The energy required for this whirlwind of activity is provided by oxidation-reduction cycles that take place within the mitochondria, the cytoplasmic corpuscles that Lynn Margullis identified in 1970 as ancient bacteria, introduced in symbiosis with the ancestors of eukaryotes and which are part of the baggage transmitted to the egg cell by the maternal oocyte.

3.3. The weight of the physical setting

From the very beginning of its construction, in the state of the original cell, the living organism adheres to the three aspects of space. The female gamete of bilateral metazoans is not isotropic. Within the cytoplasm of this voluminous cell, movements provided by the elements of the cytoskeleton create a distribution of "morphogenic" proteins and messenger RNA which leads to an axis between an "animal" pole and a "vegetative" pole, where the reserves are concentrated. This first arrangement, with respect to space with radial symmetry, constitutes a reference for the events that follow. It is the penetration of the spermatozoon that disrupts the original symmetry. The kinetic center of this micro-missile destabilizes the contractile actomyosin network of the maternal cytoplasm, whose pulses will distribute activating and inhibiting proteins. Hence the segregation of the two and then three leaves and the formation of an orifice, the blastopore. In spite of the similarity of these first stages in bilaterians, there are major differences that appear which involve the polarity of the symmetries by the mechanism of gastrulation. Indeed, the appearance of the blastopore opens up two future pathways: this orifice becomes the anterior opening of the digestive tract in annelids, roundworms (nematodes), mollusks and arthropods, which represent a majority of living forms. They are protostomal (primary mouth). In deuterostomians (secondary mouth), the blastopore becomes the posterior end of the digestive tract and the mouth later opens at the anterior end. These include echinoderms (sea urchins, starfish and sea cucumbers), and chordates (amphioxus or lancet and vertebrates). Another distinction is essential in the specification of forms. It concerns the relative position of the digestive tract and nervous system, opposing chordates to annelids and arthropods. The former are epineurian, with their nervous axis extending back and forth above the digestive tract. The latter are hyponeurian, and have an axis formed by a pair of cords that extend below the digestive tract, with the exception of the anterior pole (that of the head). Indeed, at this level, the cords form a necklace whose ganglia are the equivalent of a brain. Once it was possible to identify the genes which, through the play of cascades, are responsible for the placement of the ventral and dorsal portions of the embryo in Drosophila, the sog (short gastrulation) gene for the abdomen and Ddp (decapentaplegic) for the back, we looked for those which were also involved in this polarity in vertebrates. In the xenopus, the aquatic anuran amphibian, a model subject for the experimenters, the dorsal gene is chordin and the ventral gene is BMP4 (Brown Morphognetic Protein 4). However, analysis of the composition of these genes (their sequencing) revealed that sog is homologous to chordin, and ddp to BMP4: the gene for ventral expression in the insect is the same as that for dorsal expression in the vertebrate (see Figure 3.2). Dorsal-ventral signaling is reversed between the insect and the vertebrate, giving credence to the intuition defended by Geoffroy Saint-Hilaire against Georges Cuvier in 1830 (Le Guyader 1998).

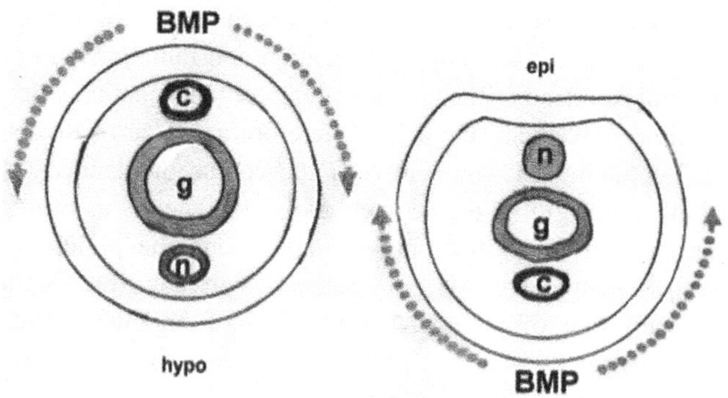

Figure 3.2. *Action sector of the ventral BMP4 gene. Insects on the left; vertebrates on the right. n: nervous system; g: digestive tract; c: circulatory system*

3.4. Mesoderm: base material for architect genes

The mesoderm is one of the main germs responsible for the form under construction. It provides the constituent materials for the organs that provide the framework and mobility of the organism. In vertebrate embryos, the mesoderm differentiates on either side of the dorsal chord, regularly producing blocks of cells, the somites, according to the rhythm of a clock (Gasc and Renous 1989; Pourquié 2001). The somites will give the future skeletal elements (vertebrae), the axial musculature, the integument. The implementation of the main lines of this architecture is dependent on a particular set of genes, among which the so-called "homeotic" genes stand out. Their discovery has upset our conception of morphogenesis by revealing a solution to the puzzle that preoccupied geneticists that were faced with the idea of an improbable combination between independent genes, leading to the construction of organs in all their complexity.

As early as the end of the 19th Century, William Bateson discovered individuals in drosophila whose appendages, antennae, legs and wings were out of place. Particularly flies with two pairs of wings instead of one, the famous bithorax model. He called this type of morphology "homeosis" or "homeotic variation". But it was only from the 1950s onwards that Edward B. Lewis gradually discovered the array of genes in Drosophila that were responsible for the establishment of the segmental compartments of the insect, with specialization corresponding to appendages. These homeotic genes have been shown to be generalized in metazoans. In bilaterals, they are arranged on the chromosomes in the same position as that of the organs whose construction they trigger along the anterior-posterior axis (Duboule and Morata 1991). This homeotic system manifests a principle of order, subject to regulation by the activation-repression process. The whole has been subjected, on the one hand, to variation, through the play of genome duplication and mutations, and on the other hand, to the natural selection of the corresponding morphofunctional devices (organs related to nutrition, movement, etc.). It is therefore a source of diversity of forms that share the fluctuating environment over the course of planetary history. This understanding of the origin of the segmentation of bilaterals answered a question raised by German naturalists and philosophers among the philosopher Friedrich Wilhelm Joseph Schelling and the poet Johann Wolfgang von Goethe, who had proposed a general rule of modular composition of organisms (Schmitt 2004). Despite the extravagances of some promoters of this theory, it has remained a typology: the existence of beings composed of units, the metamers, repeating themselves from front to back on the axis defining their bilateral symmetry. Indeed, from the Cambrian, in this astonishing explosion of metazoan forms that is revealed in the Burgess Shales dating back 500 million years (Gould 1989), segmented morphologies are dominant. However, this notion of metamerism is based on an ideal typology. Each constituent metamer is supposed to comprise a set of specialized cells which, by their association, ensure the essential vital functions: a supporting element, its muscles, its innervation, its irrigation by nutritive fluid, and even an organ of excretion. However, only annelids have this complete composition and only the mesoderm and ectoderm are susceptible to segmentation, while derivatives of the endoderm (mainly the digestive tract) are not metamerized. The mechanism of segregating cells into clone blocks, discovered in Drosophila in 1970 by Garcia-Bellido and his team, is based on surface proteins that create the recognition and adhesion of cells resulting from cloning by rhythmic mitosis. These are regulatory sequences that intervene according to a precise chronology for the positioning and shaping of segments, and thus their functionality: antenna, eye, mouth parts, legs and wings. These genetic mechanisms have been generalized to a majority of bilaterals. However, on these simple bases, the multiplicity of forms has resulted. One can think that the initial modular plan has specifically allowed a great diversity by selection, without requiring costly or impossible upheavals in the construction of the organism. Even in arthropods, where segmentation is evident, Alessandro Minelli (2003) notes a number of exceptions, disturbances in order, mergers or deletions of segments, sometimes resulting in unusual adult forms related to particular lifestyles, as in parasitic crustaceans, reduced to breeding bags. Even in the best students in the class where segmentation is evident, such as annelids and myriapods, various dispositions are observed. Some annelids produce new segments from the back of the body. Contrary to other species, which have a fixed number of segments, myriapods show both a wide range in the number of segments with locomotive appendages, and a high fixity of this number in each taxon. For Buss (1987), this diversity is a sign of the quality of a field that is subject to natural selection over the course of evolution.

Another cause of the diversification of adult forms is the delayed activation of genes during development. This is the process of heterochrony, which can, for example, produce larvae that reproduce (neotenia), thus eliminating the adult form, as in the case of the axolotl in vertebrates.

Ultimately, since these processes are dependent on a cascade of regulatory sequences that take place after the major polarity axes have been established, they are subject to epigenetic influences. This would explain the examples of fusion or simplification associated with functional specializations. Thus, the metameric pattern has opened up the possibility of a morphofunctional design from genome variations, which have a variety of sources, including epigenetic effects throughout ontogenesis (Deutsch 2012), and which are stabilized by natural selection during the diffusion of metazoans into the various environments of the planet.

3.5. Appendices and laws of mechanics

Protostomians and deuterostomians show lateral excrescences, more or less perpendicular to the main axis of the body and responding to the succession of segments. Onychophores or peripatus, discrete organisms whose organization and geographical distribution indicate their ancient character, as well as a fossil (*Ayscheiaia pedunculata*) dating from the Cambrian, have a first pair of antenna-like appendages, followed by a series of 17 to 43 pairs of soft protuberances, the lobopods, which end in claws. Their movement, coordinated according to a modulated wave train, ensures the animal's progression through the litter of tropical forests (see Figure 3.3). In annelids, the appendages or parapods are stiffened by the cuticle, which is coated with chitin, a polymer resistant to compressive forces (Wainwright *et al.* 1976). They consist of two axes in which diverticula of the general cavity extend. The body of arthropods is more or less clearly divided into regions (head, thorax, abdomen, tagma, clear in insects) and their appendages are themselves segmented.

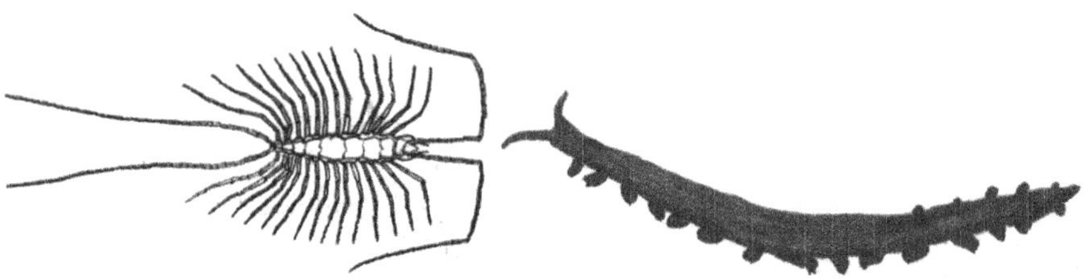

Figure 3.3. *A centipede (myriapod) and a velvet worm (onychophora)*

Since the final segmentation and the morphofunctional identity of each segment is under the control of the homeotic complex, a scenario could be proposed for the genetic events explaining the passage of an ancestor, in whom all segments would have been identical to forms showing a regional specialization, and from which even myriapods are not excluded[2]. With the appendages, segmentation is at the service of a division of labor, sensory organs, food processing, aquatic or terrestrial movement, flight or reproduction by internal fertilization, and even venomous organs of defense and attack. While ruling out final interpretations, it must be recognized that the selection of the segmental pattern in bilaterians who possess mobile appendages has played a role in their ability, in a competitive regime, to exploit the resources of new spaces, those created by geological episodes in the earth's crust. One of these episodes is well documented: the gradual establishment of stable continental margins, which allowed living beings to leave the aquatic environment in which

2 Among "millipedes", the number of segments, ranging from 191 to 325, may be fixed at birth (as in the case of scolopendras and geophiles in our gardens) or it may increase with age (as in the centipedes found in houses).

they had previously been confined. Let us forget the slogan "coming out of the water"! The expansion of living forms in continental environments actually took place over millions of years and in successive waves, primarily for ecological reasons: trophic chains had to form with organisms that could leave the aquatic environment. Between the Cambrian and the Devonian, in other words, in 130 million years, the various metazoan lines had to develop forms capable of overcoming completely new physical-chemical constraints. Life out of water requires, first of all, the periodic disappearance of the element on which most chemical reactions are based. Water becomes a limiting factor. The atmospheric environment is much more fluctuating than ocean waters: physical factors (intensity and composition of radiation) and chemical factors (gas composition, mainly nitrogen, carbon dioxide and oxygen) are subject to variations of planetary and cosmic origin. After the turbulent phases that accompanied the beginnings of biological evolution (Maurel 1994) and with the increasing intervention of living forms that only depended on water and carbon dioxide (photosynthetic organisms), the oxygen composition of the atmosphere has gradually stabilized around the current values of 20.9%. However, this element acts as the final receptor for hydrogen in dominant aerobic energy metabolisms. The advantage of life in the open air compared to the aquatic environment is therefore understandable, as water cannot dissolve more than 1% of oxygen. The continental surfaces were initially a support for cyanobacteria, capable of resisting temporary desiccation and which today still form a biofilm on the rocks of the inselbergs of Guyana, where the climatic conditions – exposure to the equatorial sun, high temperatures, torrential rain – are probably not very different from those that prevailed in the Silurian period, 400 million years ago[3]. In the beginning, this whole world was at ground level. Indeed, in the open air, gravity was exerted without restraint, unlike in the aquatic environment where it was roughly balanced by hydrostatic pressure. This is why cyanobacteria could not build the equivalent of Precambrian stromatolites on land and cnidarian polyps could not emerge to build the equivalent of gigantic coral reefs, whose fossil traces can be found in the mountains. Gravity, a vertical constraint, has two mechanical modalities to which organisms must respond structurally: a static modality and a dynamic modality (Clark 1964). First, a collapse must be resisted by incorporating materials into the structure that resist to compressive strength with minimal deformation. Living structures are, from their origin, rich in collagens, fibrous elements that are resistant to tension. Then,

3 This biofilm, sometimes soaked and gooey, sometimes dried out, is the primary producer of organic matter available to other microorganisms and the tiny arthropods and annelids that consume it (Vaçulik *et al.* 2004).

biological materials resistant to compression appear based on the principle of chain linking (polymerization) made up of molecules with six carbon atoms, whose module is glucose. For example, chitin (acetylated glucosamine) is widely used by metazoans, who share chitin with their mushroom siblings. Associated with a protein in the cuticle, it constitutes the material of an exoskeleton, a veritable armor that follows the segmentation plan of the organism, and must be renewed periodically to allow growth (molts). It can be surface tanned or enriched with calcium carbonate (as in crabs, for example), thus gaining both compression and flexion (Wainwright *et al.* 1976). Arthropods were able to cross the boundary between water and open air without difficulty thanks to this armor. Scorpions may have been among the first, as their astonishing resistance to very high-energy radiation from nuclear explosions suggests that they experienced a period when the protective ozone shield was not fully installed. Among the deuterostomes, echinoderms or ambulatory sea urchins and starfish could not have made it ashore despite their locomotor system based on the regulation of internal pressure. Moreover, their type-5 radial symmetry in adults, an innovation of life since no mineral has this symmetry deriving from a bilateral arrangement, seems more adapted to the three-dimensional aquatic environment than to the uneven surfaces of the emerged soils[4]. Among chordates, the segmented mesoderm of vertebrates that becomes rigid along the vertebral axis has allowed several modes of swimming by lateral flexion under the action of metamerized muscles, but also movement on the sea floor with the help of appendages (Renous and Gasc 2011). We can then discuss the second dynamic gravity constraint.

3.6. "Appendicular" movement on land

All organisms that combine metamerism, lateral appendages and compression-resistant materials in their bilaterally symmetrical structure were able to venture out of the water. In fact, each appendage that is rigidified by an external or internal skeleton and moved by a segmental musculature constitutes a lever capable of locally exerting an external force, from which it will receive a reaction to the surrounding environment according to the Newtonian principle of all movement, provided that it does not slip[5].

---

4 Many aquatic animals become rigid due to the internal pressure modulated by the contraction of peripheral muscles (hydrostatic skeleton). This principle could only be applicable in terrestrial environments for burrowing forms that avoid water loss. This is the case for earthworms.

5 The keratin molecule of the epidermis in various configurations ensures adhesion between the support and the soil (Gasc *et al.* 1982; Abdel-Aal 2017).

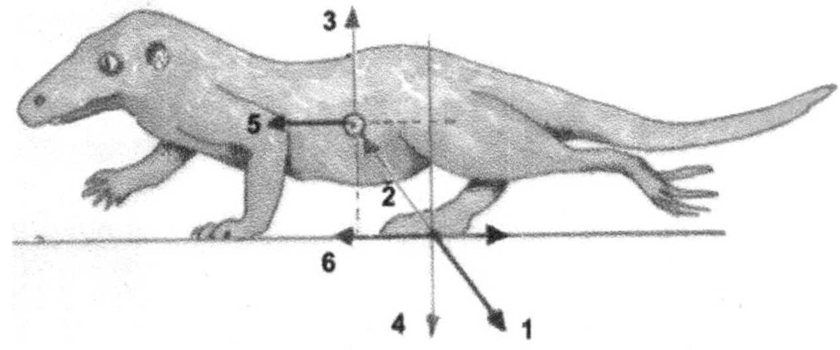

Figure 3.4. *Rear propulsion of tetrapods.1: thrust on the ground; 2: its reaction; 3: vertical component balancing the mass; 4: vertical component of the thrust; 5: horizontal driving component acting on the center of mass; 6: static frictional force ensuring adhesion. For a color version of this figure, see www.iste.co.uk/chapouthier/life.zip*

The segmental organization of the nervous system ensures the sequential action of the appendages, which are each alternately pulled backwards (support) and brought forward to a new support. The whole evolution and diversification of these animals has been based on the appearance of biomechanical devices tending towards greater efficiency, a better performance in the production of external forces and the resulting motor component which is transmitted to the center of mass of the organism. Between a cheetah and a centipede, the same fundamental laws of mechanics are applied; the only difference is that of the scale and degree of efficiency necessary for the daily life of each one (Weibel *et al.* 1998). The appearance and construction of the even appendages of bilaterians are dependent on genetic complexes, distal-less (Dll) of the Drosophila, and its counterpart Dix of the Vertebrates (Panganiban *et al.* 1997), their position and identity being dependent on the homeotic complex. However, the generality of the expression of these complexes in metazoans, including in the construction of functionally distant structures (such as gill arches and the nervous system), suggests that this genetic complex has been "co-opted" over the course of evolution. In vertebrates, the mesoderm is differentiated along the dorsal chord, a flexible axis extending from the back of the head to the posterior end. From the interaction between these three elements (chord, somites and the central nervous system), the mobilizing structures of the organism will be born: the vertebral axis, its musculature and its innervation. Like arthropods, vertebrates have articulated appendages, in other words, a series of features mobilized by muscles. This mechanism can be summed up as a set of alternating angles that form levers that offer a wide range of operating and efficiency modes. The paired limbs of vertebrates that have dispersed over the surface of continents define their quality as "tetrapods", because they have only two pairs of appendages (see Figure 3.4). However, these members in their composition and position were already present in aquatic forms among Devonian sarcopterygians (Laurin 2011).

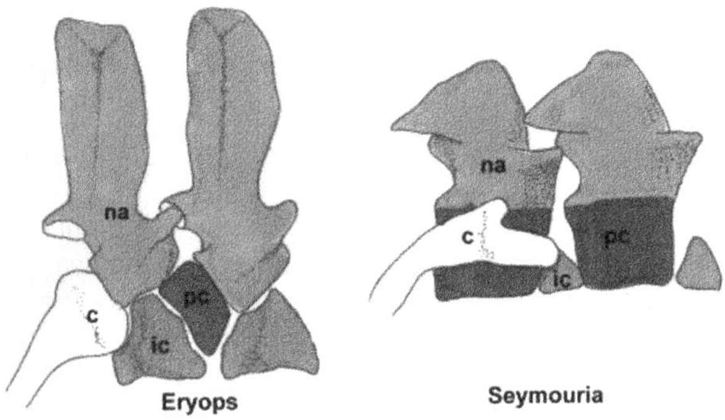

Figure 3.5. *The transition to an aerial life goes hand in hand with a consolidation of articulated vertebrae that are directly subjected to muscular forces. For a color version of this figure, see www.iste.co.uk/chapouthier/life.zip*

It is now evident that the first vertebrates on continents already possessed lungs and limbs capable of exerting locomotor forces in a gravitational regime. On the other hand, a further consolidation of the vertebrae can be observed by the fusion of three parts initially present in the aquatic ancestors (neural arch, pleurocentra and intercentra), in relation to the need to form an articulated column of units that are resistant to compression (see Figure 3.5). In terms of the limbs, the opening of the alternating angles under muscle actions determines an extension of the assembly, which exerts a force on the ground, and whose vertical component balances gravity by reaction. Thus, the anterior portion of the body can be lifted and used for the progression mechanism through the undulating action of the vertebral axis. This wave tilts the support from one side to the other. This locomotor mechanism is therefore based on a simple, rhythmic and alternating nervous control between two opposing actors, flexor and extensor, coupling the contraction wave that travels through the axial musculature and the movement of appendages. A first modification occurs when the hind limbs become attached to the vertebral axis via the pelvic girdle, whose ilium, the most dorsal element, is anchored to the sides of the vertebrae (Boisvert 2005). This ilio-sacral connection allows the transmission of the reaction of the ground to the extension of the hind limb to the body's center of mass, thanks to the rigidity of the vertebral axis which is compressed. From a forward traction, the animal has become a rear propeller, which gives it the ability to raise the center of mass located in front of the supports and to maneuver around bends.

The second change concerns the orientation of the limbs in space. The first segment, called the "stylopod", with the humerus in front and the femur in the rear, is arranged more or less perpendicular to the body axis in early tetrapods. This transverse position places the body mass in a cantilever position and a great deal of muscular energy is absorbed to lift the body mass above the ground; this is manifested by the massiveness of the bone, sculpted by the insertions of large muscles.

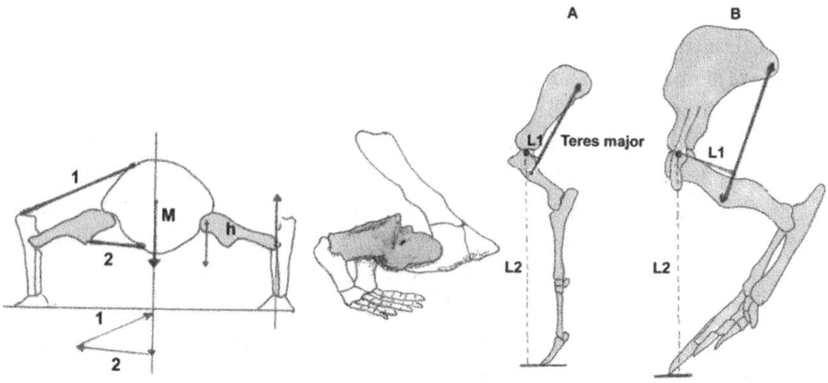

Figure 3.6. *h: transverse humerus of the first tetrapods; 1 and 2: muscles balancing mass M and parasagittal anterior limb; A: horse with a large round muscle acting in speed; B: armadillo with a muscle acting in power; L1: lever arm of the muscle; L2: resistant lever arm. For a color version of this figure, see www.iste.co.uk/chapouthier/life.zip*

From this transverse position, the stylopods move to a position that is parallel to the sagittal plane, the one passing through the axis of the body. This is the case for many sauropsids, including the hind limb of dinosaurs (and birds), as well as among the synapsids from which mammals emerge. The parasagittal limb has mechanical advantages (see Figure 3.6). All force exchanges take place in a single plane, which is that of the motor components, the lateral "parasitic" effects being balanced by the simultaneous action of the opposite side through the phases of the motor pattern. The extension acts both against gravity and in propulsion, and the bending and acceleration of the bent limb, the non-driving phase, is carried out with economy in a pendulum-like manner. The general form of the body is considerably changed by this new arrangement of limbs and has opened up a whole field of adaptations. The limbs become the main actors in movement, and their geometry in relation to the way of life reflects mechanical requirements. The relative length of the segments and the site of insertion of the muscle motors determine degrees of efficiency in specific actions: speed or power (see Figure 3.6). The parasagittal position allows the bones to resist better to compression, by increasing the diameter and particularly by the formation of columns with the vertical stylopods, and thus, very massive organisms emerge on the continents (from large dinosaurs to mastodons).

3.7. The legless

One of the surprises of evolution is the appearance, many times and very early on, of legless vertebrate forms. Carl Gans (1975) attributed the evolutionary advantage of being able to access locations that were inaccessible to those with legs, who had the same mass, to the long cylindrical form of serpents. It should be added that their mode of movement tends to suppress the energy-intensive reciprocating movement of limbs in favor of continuous gliding. Apod forms are present in all environments and their specialization can be astonishing, such as liana snakes in trees, marine snakes, burrowing snakes. An array of lissamphibians, gymnophiones, and among the squamate reptiles, amphisbaenians, snakes and more than a dozen families of lizards, all enter this pattern which is distinguished both by the loss of the locomotor role of the limbs and by an increased number of vertebrae. The vertebral axis is in fact the only movement mechanism for snake-like vertebrates and its efficiency depends on its length and the number of articulated segments. The increase in the number of articulated segments depends on a "runaway" clock that produces the somites from a mesodermal bud in the tail of the embryo. The frequency of this production rate is four times higher in snakes than in mice (Woltering 2012). The last posterior contingent of homeotic genes, Hox 13, appears to have lost its role as an end-of-process signal. In addition, all Hox genes appear to have lost their ability to regionalize somites, resulting in relative homogeneity in vertebrae form. On the other hand, the morphogenetic factor that would link somitic multiplication to limb reduction remains unknown. Is it retinoic acid? To what extent do the somitic prolongations reaching the limb outline (Raynaud and Vasse 1968) constitute this link, assuming a somitic deficiency in the induction of apical crest formation of the appendicular bud (Gasc 1970) (because the more somites produced, the poorer they are in terms of cells)? The question remains open.

3.8. And the head

Bilateral metazoans, whether visibly segmented or not, are strongly polarized by the anterior pole concentration of the organs of information and food intake, in other words, at the level of the living being's leading front in terms of the environment. According to this definition, the presence of a head is very general and this morphological feature has been called "cephalization". However, the forms and formative processes are different according to the phylum of bilaterians (Gasc 2004). In mollusks, there are two extremes: cephalopods, cuttlefish, octopus and squid, which appear to be all-headed, and lamellibranchs, oysters and mussels, which have lost their heads. The head of arthropods is complex and results from the coalescence of segments, with specialized appendages. The transformation is clearer in insects, locusts, flies or butterflies: the mouthparts can form crushing, licking or sucking tools. This is the effect of a particular program of the Hox genes complex. The case of vertebrates presents itself differently. The anterior portion of the head is formed at the front of the end of the chord, and the initial structures of the skull, the box that characterizes this line called "craniata", do not show segmentation. However, a double segmentation is apparent at the front of the hearing apparatus: ventrally at the level of the pharynx by the repetition of gill pockets (branchiomery), and dorsally at the level of the central nervous system in the future brain (rhombomery) and in the sequence of cranial nerves.

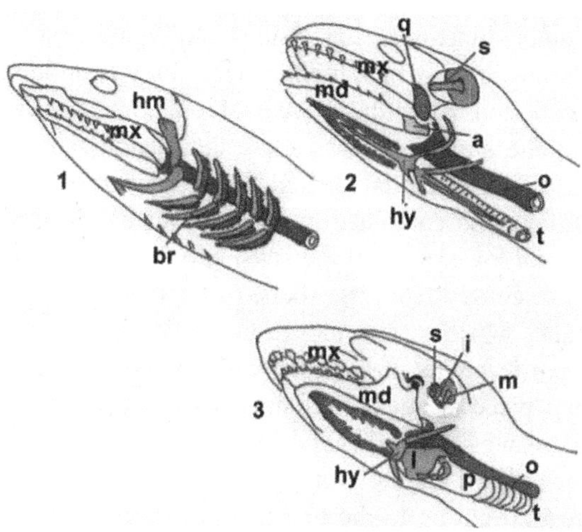

Figure 3.7. *Transformation of the hyobranchial apparatus. 1: aquatic respiration; 2: aerial tetrapod; 3: mammal; a: articular; br: gills; hm: hyomandibular; hy: hyoid; i: incus; m: malleus; o: oesophagus; q: quadrate; s: stapes; t: trachea. For a color version of this figure, see www.iste.co.uk/chapouthier/life.zip*

Functionally essential structures that have had an impact on the diversity of vertebrate forms are thus based on the principle of modules. During development, cells from the ectodermal crests of the groove forming the future nerve axis (neural crests) (Le Douarin 1982) migrate inwards along segmental paths and, in association with elements of the mesoderm, give rise to the bulk of the visceral or hyobranchial skeleton (Gasc 1967). From this material, which includes bone, muscle, blood vessels and nerves, evolution has tinkered with apparatuses specific to each of the major groups of vertebrates: the gill basket of vertebrates with aquatic respiration, the jaws and the hyoid apparatus of those that breathe ambient air. From the gill basket of the agnatha that populated the seas of the primary era, and whose only current representatives are sea hagfish and river lampreys, to the jaws of mammals adapted to various diets, this is a fantastic carnival of forms that has come about through design and a few innovations. The jaws, suspended under the skull or firmly welded to it (Schmalhausen 1924), are formed from the elements of the first segment or mandibular arch. They are innervated by the 5th cranial nerve, the trigeminal nerve. As long as vertebrates live in this environment, the following arches constitute the apparatus for capturing dissolved oxygen in the water: the gills. The transition to breathing air is accompanied by the formation of pulmonary pockets from ventral diverticula of the digestive tract, where gas exchange takes place. This reduces the role of the gill arches, an observation made in lungfish. Their material is "recycled" into other functions. Expansion into continental environments has changed the conditions under which food is taken. Prey is no longer sucked up, but is grasped and transported within the oral cavity to the intersection with the airways. This is the role of the structures that follow the mandibular arch in forming the hyoid bone and cartilage of the larynx, under the control of the 7th, 9th and 10th cranial nerves. Design also continued in the open air by the passage of the hinge, between the upper and lower parts of the jaws, into the functional sphere of hearing. Sound vibrations are concentrated by a bone rod, the stapes or stirrup, between the tympanic membrane, which is an innovation, and the oval window, which is an opening to the nerve receptors of the inner ear. Embryology showed that this bone was derived from a piece of aquatic binding material between the mandibular and hyoid (hyomandibular) arch (Schmalhausen 1924). The transformation is accentuated in the lineage of the therapsids. It has been followed in fossils for several million years, right up to the first mammals (Benton 1997). It is the entire hinge of reptilian jaws, between the joint at the bottom and the quadrate at the top, that becomes the malleus (articular) and incus (quadrate), thus completing the chain of ossicles in the middle ear (see Figure 3.7). At the same time, the musculature mobilizing the jaws is modified, changing its orientation and freeing the old hinge from mechanical stress. Complex teeth appear, allowing a gear to be produced, with a hinge constituting a new pivot (Bramble 1978). A new joint (temporomandibular) is formed between the dental bone of the mandible, forming the condyle and the squamosal bone of the skull which receives it. For obvious reasons of functionality, the two joints remained in parallel for a very long time.

Thus, deciphering the form of a living being means searching for the place it maintains in the conditions of its survival, probing the archives of its history and unravelling the mechanisms responsible for its construction. It is only in the light of this triad of research that we can hope to understand the reality of any form.

3.9. References

Abdel-Aal, H.A. (2018). Surface structure and tribology of legless squamate reptiles. *Journal of the Mechanical Behavior of Biomedical Materials*, 79, 354–398.

Benton, M.J. (1997). *Vertebrate Palaeontology*. Chapman & Hall, London.

Boisvert, C.A. (2005). The pelvic fin and girdle of Panderichthys and the origin of tetrapod locomotion. *Nature*, 438, 1145–1147.

Bramble, D.M. (1978). Origin of the mammalian feeding complex: Models and mechanisms. *Paleobiology*, 4, 271–301.

Bucher, D., Anderson, P.A.V. (2015). Evolution of the first nervous system: What can we surmise? *The Journal of Experimental Biology*, 216, 501–503.

Buss, L.W. (1987). *The Evolution of Individuality*. Princeton University Press, Princeton.

Clark, R.B. (1964). *Dynamics in Metazoan Evolution*. Oxford University Press, Oxford.

Cohn, M.J., Tickle, C. (1999). Developmental basis of limblessness and axial patterning in snakes. *Nature*, 399, 474–479.

Deutsch, J. (2012). *Le gène: Un concept en évolution*. Le Seuil, Paris.

Duboule, D., Morata, G. (1991). Colinearity and functional hierarchy among genes of the homeotic complexes. *Trends Genet.*, 10, 358–364.

Focillon, H. (1996). *Vie des formes*, 6th edition. PUF, Paris.

Foucault, M. (1966). *Les mots et les choses*. Gallimard, Paris.

Gans, C. (1975). Tetrapod limblessness: evolution and functional corollaries. *Am. Zool.*, 15, 455–467.

Gasc, J.-P. (1967). Squelette Hyobranchial. In *Traité de zoologie*, volume 16, Grassé P.-P. (ed.). Masson, Paris, 550–583.

Gasc, J.-P. (1970). Réflexions sur le concept de “régression” des organes. *Revue des questions scientifiques*, 141, 2, 175–195.

Gasc, J.-P. (2004). *Histoire naturelle de la tête. Leçon d'anatomie comparée*. Vuibert, Paris.

Gasc, J.-P. (2006). Form, function, transformation. *C. R. Palevol*, 5, 489–498.

Gasc, J.-P. (2013). Le dilemme de Cuvier: reconstituer les histoires de vie ou découvrir l'histoire de la vie. *Revue de Paléobiologie*, 32, 2, 3303–314.

Gasc, J.-P., Renous, S. (1989). Les données ontogéniques peuvent-elles expliquer l'origine évolutive des squamades apodes? In *Ontogénèse et évolution. Colloque international CNRS, Geobios*. Lyon, 173–186.

Gasc, J.-P., Renous, S., Diop, A. (1982). Structure microscopique de l'épiderme palmaire du saurien coleodactylus amazonicus (Andersson 1918) (sphaerodactylinae), comparée à celle de l'épiderme des feuilles de la litière, substrat locomoteur de l'animal. *Comptes Rendus – Académie Sciences Paris*, 294, 169–174.

Gayon, J. (2005). De la biologie comme science historique. *Les Temps modernes*, 630–631, 55–67.

Gould, S.J. (1989). *Wonderful Life*. Norton and Company, London/New York.

Haeckel, E. (1866). *Generelle Morphologie der Organismen: allgemeine Grundzüge der Organismen Formen-Wissenschaft, mechanisch begründet durch die von Charles Darwin reformite Descendent Theorie*. G. Reimer, Berlin.

Haeckel, E. (1898). *Kunstformen der Natur*. Bibliographisches Institut, Leipzig.

Kardong, K.V. (1998). *Vertebrates, Comparative Anatomy, Function, Evolution*. WCB McGraw-Hill, Boston.

Laurin, M. (2011). Paleontological evidence: origin and early evolution of limbed vertebrates. In *How Vertebrates Moved onto Land*, Bels, V., Casinos, A., Davenport, J., Gasc, J.-P., Jamon, M., Laurin, M., Renous, S. (eds). Publications scientifiques du Muséum, Paris.

Le Douarin, N. (1982). *The Neural Crest*. Cambridge University Press, Cambridge.

Le Guyader, H. (1998). *Geoffroy Saint-Hilaire*. Belin, Paris.

Maurel, M.C. (1994). *Les origines de la vie*. Syros, Paris.

Minelli, A. (2003). *The Development of Animal Form*. Cambridge University Press, Cambridge.

Panganiban, G., Irvine, S.M., Lowes, C., Roehl, H., Corley, L.S., Sherdon, B., Grenier, J.K., Fallon, J.F., Kimble, J., Walker, M., Wray, G.A., Swalla, B.J., Martindale, M.O., Carroll, S.B. (1997). The origin and evolution of animal appendages. *Proc. Natl. Sci. USA*, 94, 5162–5166.

Pourquié, O. (2001). The vertebrate segmentation clock. *Journal of Anatomy*, 169–175.

Raynaud, A., Vasse, J. (1968). Les relations entre les somites et les ébauches des membres antérieurs, chez l'embryon d'orvet (anguis fragilis L.). *Archives. Anatomie, microscopique. Morphologie Expérimentale*, 57, 227–254.

Renous, S., Gasc, J.P. (2011) Terrestrial locomotion of lower extant vertebrates. In *How Vertebrates Moved onto Land*, Bels, V., Casinos, C., Davenport, J., Gasc, J.-P., Jamon, M., Laurin, M., Renous, S. (eds). Publications scientifiques du Muséum, Paris, 129–170.

Schmalhausen, I.I. (1924). *Origine des vertébrés terrestres*. Nauka, Moscow.

Schmitt, S. (2004). *Histoire d'une question anatomique : la répétition des parties*. Muséum national d'Histoire naturelle, Paris.

Vaçulik, A., Kounda-Kiki, C., Sarthou, C., Ponge, J.-F. (2004). Soil invertebrate activity in biological crusts on tropical inselbergs. *European Journal Soil Science*, 55, 539–549.

Wada, H., Garcia-Fernandez, J., Holland, P.W.H. (1999). Colinear and segmented expression of amphioxus Hox genes. *Developmental Biology*, 213, 131–141.

Wagner, G.P. (1996). Homologues, natural kinds and the evolution of modularity. *American Zoologist*, 36, 36–43.

Wainwright, S.A., Biggs, W.D., Currey, J.D., Gosline, J.M. (1976). *Mechanical Design in Organisms*. Princeton University Press, Princeton.

Weibel, E.R., Taylor, C.R., Bolis, L. (1998). *Principles of Animal Design*. Cambridge University Press, Cambridge.

Woltering, J.M. (2012). From lizard to snake; behind the evolution of an extreme body plan. *Current Genomics*, 13, 289–299.

4

On Growth and Form: Context and Purpose

Jean-Pierre GASC
National Museum of Natural History, Paris, France

Published in 1917, and then in highly-expanded version in 1942, *On Growth and Form* is a book written by Scottish zoologist D'Arcy Wenworth Thompson, who was a professor at the University of St Andrews, and holds an almost legendary place in the scientific literature, and even beyond. As in all legends, it has inspired people in every era, as shown by the various commentators who have made it out to be the work of a mathematician, a lone heretic, or an amateur against the modern aspects of biology. But what place did D'Arcy Thompson really hold in the scientific community in which his long life (1860–1948) took place? In order to judge it, as well as evaluate its influence until the 21st Century, it is necessary to refer to the sources and go beyond the simple overview of this popular book. One will be astonished when lured into a historical examination of the debates that have marked the life sciences at least since the beginning of the 19th Century, and especially since the publication of Charles Darwin's first book in 1859. It is from the elements of Thompson's life, his professional career, and his working relationships that it is possible to understand this intellectual plunge, the concepts discussed, and to trace the maturation of his ideas in agreement or disagreement with the protagonists of the adventure – that of the process of consolidating a theory of evolution. He was 57 years old when he published this work in 1917. He already had a long career behind him: 32 years as Professor of Biology at the University College of Dundee, where he created a museum of zoology, following which he was appointed Chair of Zoology at the University of St Andrews. He is known for his

The Explosion of Life Forms,
coordinated by Georges CHAPOUTHIER and Marie-Christine MAUREL. © ISTE Ltd 2020.

translation of Aristotle's *History of Animals* in 1910, but equally for his translation of Herman Müller's *Die Beruchtung der Blumen durch Insecten* in 1883, which Charles Darwin wrote the preface to in a work that founded the concept of co-evolution. He was also an expert in marine biology and an advocate for endangered species, particularly the marine otter fur hunt in the Bering Strait. In his museum, he had collected a skeleton of the Steller's sea cow, a giant sirenian that had recently disappeared from those areas. In Dundee, he was an administrator of the hospital and founded an association for the housing of poor families. Lastly, he had a passion for mathematics, which he practiced as an amateur.

Peter Medawar has devoted several pages to *On Growth and Form* (1984). He underlines its originality, both in style and language (which he describes as: "I think that *Growth and Form* is beyond comparison the finest work of literature in all the annals of science that have been recorded in the English tongue"), and in ideas, which he finds timeless. And yet, this landmark must be placed on the movement of ideas that agitated scientific and philosophical circles between 1880 and 1930, regarding the theory of evolution. Espinoso (2014) consulted D'Arcy Thompson's correspondence, which allowed him to appreciate the importance of his relationships with some of the key players in the biology community of his time. Thompson belonged to the generation of British students who were immersed in the debates that followed the first steps of Darwinism. Comparative anatomy and embryology dominated then, and were brought to the heart of the debates on phylogeny by Thomas Henry Huxley, Ernst Haeckel, Karl Gegenbaur and Oscar Hertwig (Gasc 1996a). It was also the period of discovery of the marine world and new models of study in newly created research stations. The mechanism of fertilization is depicted in algae, the first cellular phases of embryonic development in sea urchins. His translation of Müller's book on pollination in 1883 shows his involvement in the ongoing debates from his youth, and he participated in the development of marine biology through numerous publications on organisms whose diversity was being discovered. German universities and laboratories have been an international magnet for English students. Anton Dohrn (1840–1909), a graduate of the University of Jena where he was a pupil of Haeckel, founded the Stazione zoologica di Napoli (Zoological Station of Naples) in 1872. This experimental research station is part of a network project (inspired by the railway model!) that is international. It has played an essential role in understanding embryonic development, using marine organisms as models. The "Vivarium" in Vienna (Biologische Versuchsanstall) was founded in 1902 by Hans Leo Przibram (1874–1944). It is an international laboratory which allowed work in "experimental zoology", where biological and physical-chemical approaches were combined, following on from the work of Berthold Hatschek (1854–1941). Best known for his work on the embryology of the Amphioxus and his hypotheses on the origin of bilaterians, Hatschek was interested in the physical-chemical properties of "colloids". He obtained *mimics* of organic forms, jellyfish and polyps, from mixtures of substances. Thompson's correspondence preserved at the University of St Andrews reveals the close and lasting relationship between him, Hatschek and Przibram. Most of the experimental examples cited in *On Growth and Form* are the result of these exchanges, because he was not an experimentalist. Müller (2017) cites this situation described by the mathematician and historian Marjorie Senechal:

> Where D'Arcy Thompson saw analogues, the Przibram brothers did experiments. Where he found elegance and simplicity, they found chaos and complexity. Where they glimpsed biological laws, he suspected leaps of imagination. But [he] and the Przibram were allies in the nascent international campaign to infuse biology with physics and chemistry.

Indeed, the Vienna Vivarium was the international crucible of a trend of experimental biology, directly derived from the comparative anatomy and embryology of the 19th Century, which, fertilized by evolutionary ideas, would focus on the conditions in which the early construction of organisms took place. This trend also includes the influence of Wilhelm Roux (1850–1924), again a pupil of Haeckel, whose *Entwickelungsmechanik* ("mechanics of development") opened up the confluence of heredity, environmental factors and functional activity. This is obviously a field that does not fit in with the dominant developments in biology following Weismann, de Vries and Morgan. Indeed, at the end of the 19th Century, morphologists were mainly concerned with phylogeny, the relationships of filiation between large groups of multicellular organisms. Haeckel's influence was noticeable, even if his "fundamental biogenetic law" of recapitulation, which nested phylogenesis within ontogenesis, staggered in the face of the number of exceptions discovered by the work of embryologists. Under the banner of Darwinism and with great talent, Haeckel defended a syncretic doctrine in which a trace of pantheism from the *Naturphilosophie* of the German romantics and a reference to Lamarck coexisted. However, in 1885, August Weismann (1834–1914) closed the door to the notion of a heredity of traits acquired in the course of individual life. He laid down the rule that organisms are composed of two independent cellular compartments. On the one side, the "germ plasma" consisting of the gametes that ensure generational transmission, resulting from a particular mode of division described by Hertwig, and on the other side, the somatic cells that ensure daily functioning and disappear when they die. This is a real conceptual revolution in the life sciences (Buss 1987). Until then, "soft" heredity, associated with Lamarck, as well as the "use and no-use" evolutionary rule, was accepted by all, including Darwin, as a mechanism for the transmission of traits and a source of transformation. Shortly afterwards, Hugo de Vries (1848–1935) rediscovered Mendel's laws, which demonstrate the independence of "factors" in hereditary transmission, and revealed the frequency of discontinuous variations, which he called "mutations". It was then that with the experimental program of the Thomas Hunt Morgan School (1866–1945), adopting Drosophila as a model, a new discipline was born: genetics. At first, it focused on the mechanisms of hereditary transmission through the interplay of corpuscles, the material support of Mendel's "factors" that Johannsen called "genes" in 1909 and which are located in chromosomes.

Most morphologists will refuse to fall by the wayside, especially embryologists, who chart a distinct course, leaving phylogenetic considerations to paleontologists and concentrating on experimental analysis of the sequence of events during embryogenesis. In short, coming back to the original mind of Karl von Baer. Biology is no longer a unified science and, while the theory of evolution is part of the conceptual landscape, the mechanisms proposed by Darwin are sidelined, especially the role of chance in variation and the intervention of natural selection. This is "the eclipse of Darwinism" (Bowler 1983). In this context, D'Arcy Thompson takes a distant, even distrustful attitude towards what he calls "modern biologists", and pursues an original program.

4.1. D'Arcy Thompson's program

As original as it is, this program is nevertheless part of a school of thought that wanted to refute the division of sciences exposed by Immanuel Kant. Indeed, according to Kant, on the one hand there are the "Newtonian" sciences, where universal laws based on the quantification of phenomena apply, and on the other hand, there are sciences such as those dealing with the living world, where finality dominates and whose complexity would prevent the application of quantification procedures. Some of Kant's German pupils were guided, following Schelling, to promote a "Naturphilosophy" which sought laws specific to living beings, that were also universal and even cosmic. This was the breeding ground for a vitalist movement and attempts to introduce a mathematical approach referring to Pythagoras, such as among the anatomists Oken and Carus, led to speculations bordering on the mystical. At the same time, and also influenced by Kant, Georges Cuvier laid down the principles of a correlation between the organs involved in vital functions. Thus, in his search for Kantian criteria, he did not hesitate to compare the relationships he observed in mammals to a mathematical equation, such as those between diet, the shape of the mandibular condyle, the type of teeth, the composition of the digestive tract and the shape of the tip of the legs (1812). Founding comparative biology, he states his ambition: "whoever would rationally possess the laws of organic economics could remake the whole animal." It is also with ambition that D'Arcy Thompson begins his work with a long introduction, a form of epistemological reflection that allows him to situate this book in the history of life sciences and its relationship with philosophy. He summarizes his program in these words:

> We want to see how, in some cases at least, the forms of living things, and of the parts of living things, can be explained by physical considerations, and to realize that in general no organic forms exist save such as are in conformity with physical and mathematical laws. And while growth is a somewhat vague word for a complex matter, which may depend on various things, from simple imbibition of water to the complicated results of the chemistry of nutrition, it deserves to be studied in relation to form: whether it proceed by simple increase of size without obvious alteration of form, or whether it so proceed as to bring about a gradual change of form and the slow development of a more or less complicated structure.

Everything is said in this passage, with the inimitable literary style of this author. He then develops this program at the various levels of integration of the organization, considering the dominant forces for each of them. At the cellular level, it is the surface tension, with the effects of surface revolution and the conditions of stability, which leads to a limited number of forms. The association of tissue cells and the conditions of stability of the forces present obey the same constraints of surface energy, and the state of equilibrium leads to the hexagonal shapes so generally widespread. Thompson cites a long passage by Buffon in this connection which, in his *Natural History*, already mechanically justified the generality of this symmetry that makes it possible to divide a space by limiting the contact surfaces. At the level of organs, he examines the case of the blood vessel connection system, based on the hypothesis of a minimum energy cost for the cardiac muscle in the face of the need to deliver oxygen to the whole organism. In particular, he is interested in the relationships between the cross-sectional areas at successive levels of the vessel connections. As the resistant friction increases as the diameter decreases, it is expected that the sum of the capillary areas will exceed the area of the original trunks, if our reasoning is it is to limit cardiac expenditure. It should be noted that, on this subject, D'Arcy approaches the concept of symmorphosis, later developed in physiology by Weibel *et al.* (Gasc 1999), in search of an optimization of structural-functional relations; however, these authors have a Darwinian perspective of selective advantage, which he rejects.

Support structures (skeletons of single-cells, shells and dynamic vertebrate architectures) are widely developed. In this field, Thompson is confronted with the complex problem of the chemistry of associations between "colloids" and crystalline forms of calcium salts, carbonates and silicates. In his time, research in biomineralization was only at the stage of describing the structures, and Haeckel's theories on the genesis of these formations and the position of mineral spicules between the cells of sponges by an evolutionary phenomenon of selection seem, to him, to be refuted insofar as similar forms are found in distant beings in the classification, and therefore, according to Haeckel, without a close ascendant. The finesse of his argumentation can be seen in his position on the very ground of the theory which he declares not to be adhering to, revealing the trap of what are today called homoplasias, and questioning the a priori phylogenetic value of any trait. He is clearly convinced that physical forces are more important than phylogenesis in determining the form of organisms. Thus, by analyzing the spiral forms of mollusks, he indicates that the simple speed of the generating curve during growth determines the difference in shell forms. However, it is with regard to the variations in the form of radiolarians, which can be followed over millions of years, that he would be more explicit in his rejection of a systematic application of the concept of evolution over time:

> … the question stares us in the face whether this be an "evolution" which we have any right to correlate with historic *time*. The mathematician can trace one conic section into another, and "evolve" for example, through innumerable graded ellipses, the circle from the straight line; which tracing of continuous steps is a true "evolution", though time has no part therein.

Thompson thus opposes two perspectives of explanation of forms, that of the physicist and that of the biologist:

> In the order of physical and mathematical complexity there is no question of the sequence of historic time. The forces that bring about the sphere, the cylinder or the ellipsoid are the same yesterday and tomorrow.

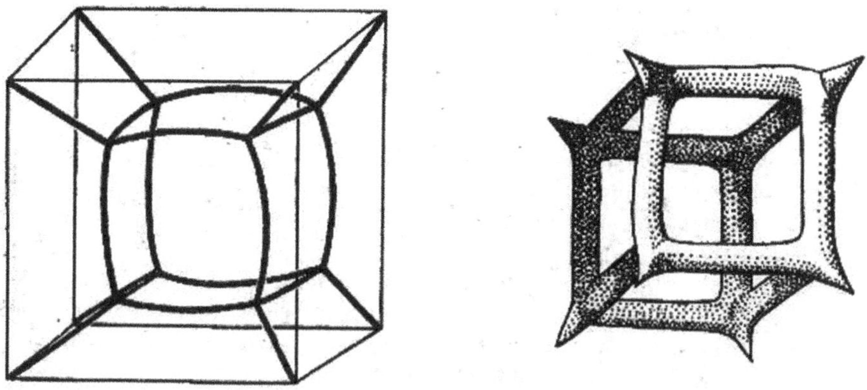

a) b)

Figure 4.1. *Experimental creation of analogous forms in* On Growth and Form. *a) Bubble enclosed in a cubic cage; b) skeleton of a radiolarian*

However, Thompson thus denounced the confusion made by many evolutionists of his time between morphological series and evolutionary series, and he questioned the role of a natural selection of the "most capable" forms:

> That things not only alter but improve is an article of faith, and the boldest of evolutionary conceptions. How far it be true were very hard to say; but I for one imagine that a pterodactyl flew no less well than does an albatross, and that Old Red Sandstone fishes swam as well and easily as the fishes of our own seas.

With the chapter entitled "On Form and Mechanical Efficiency", he analyzes examples of the relationship between form and function; in other words, he addresses the delicate issue of adaptation. Referring to the techniques of the engineer and the experience of builders, he is then in line with biomechanics, from Borelli's *De Motu Animalium* (1685), and his contemporaries, with the work of Georg Hermann von Meyer (1815–1892) and Julius Wolff (1836–1902), who interpreted the orientation of the trabeculae of bone tissue in human beings in terms of the mechanical strains in relation to the erected station. However, and contrary to Wolff, who placed his famous "law" of minimum bone substance for optimum efficiency in the Darwinian perspective of selective advantage, he rejected any genetic influence in processes where form seems to be congruent with function:

> In the biological aspect of the case, we must always remember that our bone is not only a living, but a highly plastic structure; the little trabeculae are constantly being formed and deformed, demolished and formed anew. Here, for once, it is safe to say that "heredity" need not and cannot be invoked to account for the configuration and arrangement of the trabeculae: for we can see them at any time of life in the making, under the direct action and control of the forces to which the system is exposed.

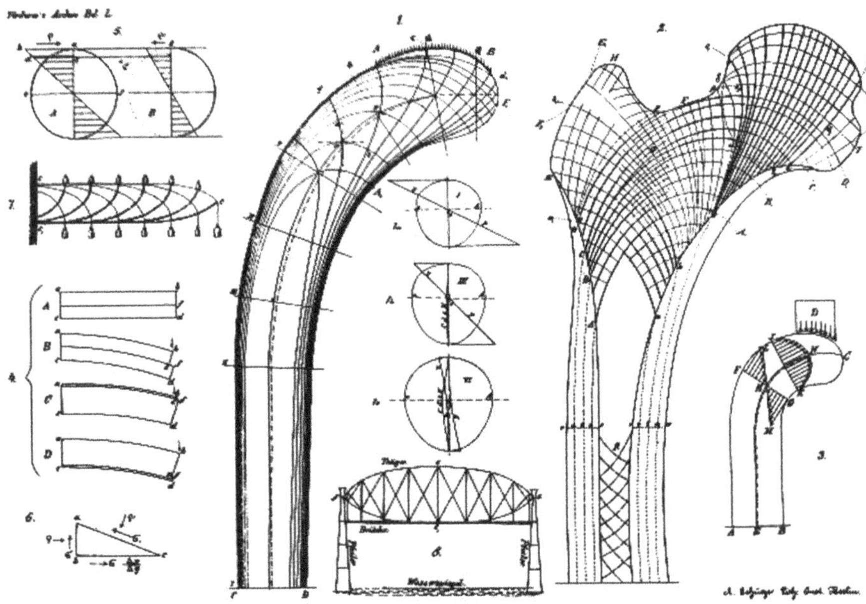

Figure 4.2. *Illustration of "Julius Wolff's Law" (1892), always in the spotlight in orthopedic surgery*

According to Thompson, therefore, the adequacy between form and function results from an adaptation of the organism to immediate physical conditions and not from an adaptation in the evolutionary sense. As for the skeletal framework as a whole, comparison with buildings, in particular with the various categories of bridges, enables him to highlight the need to consider it above all as an assembly of bony parts, tendons and muscles mobilized in actions. But, above all, he insists on the difficulty of moving from a static vision, that of buildings, to an understanding of the dynamic conditions in which this osteomuscular system intervenes in movement. Why did he overlook the French school of biological physics (1901), led by the engineer Gariel, the veterinarian Chauveau, the physicist d'Arsonval and the physiologist Marey? It can also be noted that the program he presents leads to functional morphology, through the steps of Hans Böker (1935–1937), James Gray (1944) and McNeil Alexander (1968). This injunction to consider the organism as a whole while immersed in a system of forces led him to, on the one hand, adopt a holistic perspective, and on the other, to call upon a Lamarckian mechanism. Critical of the excesses of the phylogeneticists of his time, mainly paleontologists such as Henry Fairfield Osborn (1857–1935), who promoted orthogenesis and rightly denounced syllogisms and the use of isolated traits as phylogenetic markers, Thompson remained skeptical of the theory of evolution. This reserve would be shared by many biomechanicians (Gasc 2006). Nevertheless, the Russian school, little known because it was a victim of the obligation not to publish outside the USSR during the Stalinist period, developed an "evolutionary morphology", in line with Alexei Nicolaevitch Severtsov's thought (1866–1936), through a series of comparative studies of adaptations of the locomotor and manducatory apparatus, where the rules of rational mechanics are applied (Yudin 1957; Dzerzhinsky 1972; Gambaryan 1960). The 1974 pirated translation from Russian into English of *How Mammals Run: Anatomical Adaptations* by Peter Pavel Gambaryan, the French-Russian collaboration on bird beaks (Korzun *et al.* 2004), and the recent work on the origin of flight in mammals (Panyutina *et al.* 2015) demonstrate the effectiveness of biomechanics in the framework of the Darwinian theory of evolution.

4.2. Application of mathematics to morphometry

The best-recognized contribution of *On Growth and Form* is contained in the chapter entitled "On the Theory of Transformation, or the Comparison of Related Forms". It is indeed in these pages that D'Arcy Thompson conducts his reflections on the use of mathematics in the definition, description and thus comparison of living forms. His reasoning is inspired by the adage attributed to Galileo Galilei: "the book of Nature is written in the characters of Geometry", but also by the definition of form enunciated by Aristotle, which he had translated in his youth and for whom form takes precedence over matter. The starting point is on the field of comparison. In the search for recognition and classification, the difference between two forms is expressed in terms of proportion, of relative size. The Cartesian coordinate method, which characterizes the form of a mathematical curve, and therefore its equation, is then imposed. On the condition that the comparison is made between related beings, in other words, those that are homogeneous in their composition, by using invariant landmarks, it not only becomes possible to replace a description by a quantified definition, but to also highlight the sites of deformation, to put oneself on the path of the cause of transformations, thus returning to Goethe's initial definition of morphology, in search of "metamorphoses". The figures that Thompson produced, which have become so classic, reminiscent of the anamorphoses of painters, are applied to various groups of present-day and fossil beings, and even to the various forms of leaves and their arrangement along the stem (phyllotaxis). Taking the example of the pelvis in the lineage of birds, it calculates intermediate configurations and reveals the remarkable stability of the general form. It also shows the ability of this method to remove a fossil from a presumed continuous lineage over time when anisotropy occurs.

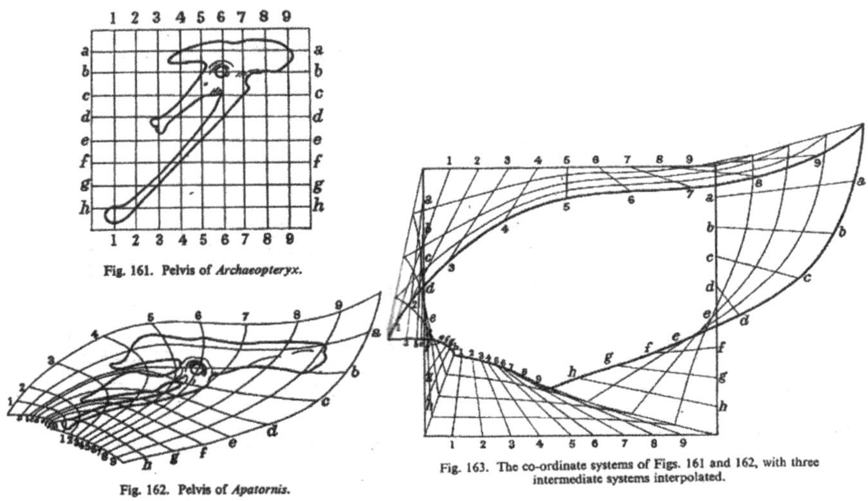

Figure 4.3. *Bird pelvic passage analysis in* On Growth and Form

Of course, for Thompson, this method essentially shows the direction of the forces that are responsible for changes in form; in other words, it indicates the relationship with function. However, he is aware that his approach is limited to an outline, to a projection on a plane, without integration of the third dimension. To present the method, he chooses fish, because their silhouette alone is a source of information for their recognition and the study of their growth.

It is precisely this zoological example that was at the origin of a veritable explosion of work in the new field of "geometric morphometry", long after his death (Bookstein *et al.* 1985). In 1988, at the congress of the American Society of Ichthyology in Ann Arbor, University of Michigan, a workshop was held that revealed the resources of computer applications in form analysis. The presence of many zoologists and paleontologists from around the world at this workshop led to the rapid spread and further development of these methods, of which D'Arcy

Thompson was the first designer (Bookstein 1982). Researchers now have the tools they lacked: taking three-dimensional coordinates, modeling the image and immediately processing a very large amount of data, thus making statistical control possible (Rohlf and Marcus 1993). The field of applications of geometric morphometry is very broad, from the study of fossils to expression recognition (Baylac *et al.* 2003; Gielis 2003; Mardia *et al.* 2018). In a critical review that is unfortunately limited to American literature, Cooke and Terhune (2015) rightly point out that it is a complementary technique, in particular to studies of functional morphology.

However, it would be wrong to limit D'Arcy Thompson to this illusory pioneer role of a method and to limit oneself to quoting his work in the last chapter! Just as Darwin's thought is often caricatured because it is quoted without having been read, *On Growth and Form* deserves special attention. In its profusion and despite the inevitable obsolete passages, it is a major and coherent work. In particular, it contains a profound reflection on scientific research involving the unity of the sciences, which was against the trend in the early years of the 20th Century, a time of babelization resulting from the specialization of disciplines. On the other hand, a century later, it appears to be on the right track. His positions on the theory of evolution and early genetics must be seen in the light of a fiercely independent spirit that was irritated by the dogmatic positions and assertions expressed in these fields by many of his contemporaries. On the other hand, it is surprising that he did not grasp the scope of the application of mathematics to the study of evolutionary processes, a movement which developed from the 1920s onwards in the study of hereditary transmission and natural selection in the extension of genetics to populations, since he did not mention it in the revised 1942 edition. Similarly, the advances in biochemistry that would be decisive in evolutionary biology did not attract his attention. In fact, his mind was entirely turned towards the geometrization of patterns and he deliberately limited the study of processes to the action of physical forces that were external to the work in the elaboration of forms.

4.3. References

Alexander, R.McN. (1968). *Animal Mechanics*. Sidwick & Jackson, Oxford.

Baylac, M., Villemant, C., Simbolotti, G. (2003). Geometric morphometrics, pattern recognition and species complexes. *Biological Journal of the Linnean Society*, 80, 89–98.

Böker, H.H. (1935). *Einfürung in die vergleichende biologische Anatomie der Wirbeltiere*, volume 1. Gustav Fischer, Jena.

Böker, H.H. (1937). *Einfürung in die vergleichende biologische Anatomie der Wirbeltiere*, volume 2. Gustav Fischer, Jena.

Bookstein, F.L. (1982). Foundations of Morphometrics. *Ann. Rev. Ecol. Syst.*, 13, 451–470.

Bookstein, F.L., Chernoff, B., Elder, R.L., Humphries, J.M. Jr., Smith, G.R., Strauss, R.E. (1985). *Morphometrics in Evolutionary Biology*. Academy of Natural Sciences Press, Philadelphia.

Borelli, J.A. (1685). *De Motu Animalium*, 2nd edition. Petrum Vander, Lugduni Batavorum.

Bowler, P.J. (1983). *The Eclipse of Darwinism*. The Johns Hopkins University Press, Baltimore.

Buss, L.W. (1987). *The Evolution of Individuality*. Princeton University Press, Princeton.

Cooke, S.B., Terhune, C.E. (2015). Form, function, and geometric morphometrics. *Anat. Rec.*, 298, 5–28.

Cuvier, G. (1812). *Recherches sur les ossements fossiles de quadrupèdes. Discours préliminaire*. Deterville, Paris.

D'Arsonval, A., Chauveau, A., Gariel, C.M., Marey, E.J. (eds) (1901). *Traité de physique biologique*, volume 1. Masson, Paris.

Dzerzhinsky, F.Y. (1972). *Biomechanical Analysis of the Yaw Apparatus of Birds*. Moscow University Publications, Moscow.

Espinoso, M. (2014). Problematic "idiosyncrasies": Rediscovering the historical context of D'Arcy Wentworth Thompson's science of form. *Science in Context*, 27(1), 79–107.

Gambaryan, P.P. (1960). *Adaptive Features of the Locomotory Organs in Fossorial Mammals*. Izdatelstvo Akademii Nauk Armyanskoi SSR, Erevan.

Gambaryan, P.P. (1974). *How Mammals Run*. John Wiley and Sons, New York.

Gasc, J.-P. (1996a). Anatomie comparée. In *Dictionnaire du darwinisme et de l'évolution*, Tort, P. (ed.). PUF, Paris, 78–85.

Gasc, J.-P. (1996b). Évolution des organismes. In *Dictionnaire du darwinisme et de l'évolution*, Tort, P. (ed.). PUF, Paris, 1506–1529.

Gasc, J.-P. (1999). Analyse d'ouvrage: E.R. Weibel, C.R. Taylor, L. Bolis (eds). Principles of animal design. The optimization and symmorphosis debate. *Revue d'écologie (La terre et la vie)*, 554(1), 92–96.

Gasc, J.-P. (2006). Form, function, transformation. *C.R. Palevol*, 5, 489–498.

Gielis, J. (2003). A generic geometric transformation that unifies a wide range of natural and abstract shapes. *American Journal of Botany*, 90(3), 333–338.

Korzun, L.P., Erard, C., Gasc, J.-P., Dzerzhinsky, F.J. (2004). Le bec des oiseaux ou quand l'anatomiste et l'étho-écologiste ont besoin l'un de l'autre. *Alauda*, 72, 4, 259–280.

Mardia, K.V., Bookstein, F.L., Khambay, B., Kent, J.T. (2018). Deformations and smile: 100 years of d'Arcy Thompson's *On Growth and Form*. *Significance*, 15(3), 20–25.

Medawar, P. (1984). *Pluto's Republic*. Oxford University Press, Oxford.

Müller, G.B. (ed.). (2017). Biologische Versuchsanstall: An Experiment in the Experimental Sciences. In *Vivarium: Experimental, Quantitative, and Theoretical Biology at Vienna's Biologische Versuchsanstall*. MIT Press, Cambridge.

Panyutina, A.A., Korzun, L.P., Kuznetsov, A.N. (2015). *Flight of Mammals: From Terrestrial Limbs to Wings*. Springer, Berlin.

Rohlf, R.J., Bookstein, F.L. (1990). *Proceedings of the Michigan Morphometrics Workshop*. University of Michigan Museum of Zoology, Ann Arbor.

Rohlf, R.J., Marcus, L.F. (1993). A revolution morphometrics. *Trends in Ecology and Evolution*, 8(4), 129–132.

Thompson, D'A.W. (1915). Morphology and mathematics. *Trans. R. Soc. Edinburgh*, 50, 857–895.

Thompson, D'A.W. (1917). *On Growth and Form*. Cambridge University Press, Cambridge.

Wolff, J. (1869). Ueber die Bedeutung des Architectur der Spongiösen Substanz. *Centralblatt für die medizinischen Wissenschaften*, 54, 949–951.

Yudin, K.A. (1957). On problems of morphological studies in ornithology. *Zoolojiskii Zhurnal*, 36(1), 113–121.

5

The Emergence of Form in the History of Epigenetics

Jonathan B. WEITZMAN
CNRS, University of Paris, France

5.1. Introduction

When Conrad Waddington coined the word *epigenetics* half a century ago, he had a cheeky word-play in mind. On the one hand, *epigenetics* was related to *genetics* and the need to describe biological events above or beyond genetic mechanisms. On the other hand, the word was inspired by the philosophical term *epigenesis*, the emergence of form through a series of steps. Waddington was interested in how genes contribute to the generation of form in the developing embryo. Today, we know much about the epigenetic mechanisms by which environmental cues and genetic features are integrated and contribute to beautiful forms and shapes underlying Waddington's *epigenetic landscape*.

I envisage a scene in which the embryologist Conrad Hal Waddington is sitting atop a mountain, looking out over the city of Edinburgh, contemplating form. Perhaps he is sitting on the mythic Arthur's Seat, an extinct volcano described by novelist Robert Louis Stevenson as "a hill for magnitude, a mountain in virtue of its bold design". This is mere speculation. But, as Waddington gazes out over the Scottish mountains and the city that houses the Epigenetics Research Group that he founded, he no doubt contemplates the origins of form and the impact of structure. For Waddington was fascinated by form for most of his career. These thoughts led him to coin the term *epigenetics* to describe the study of the emergence of biological form and structure. Waddington's most well-known legacy is the metaphor of a

The Explosion of Life Forms,
coordinated by Georges CHAPOUTHIER and Marie-Christine MAUREL. © ISTE Ltd 2020.

mountain, the concept of the *epigenetic landscape* that took form itself during his years of writing and left a framework to think about developmental biology and genetics. Although the concept of epigenetics has adopted many different *forms* with an *explosion* over the last few decades (Goldberg *et al.* 2007; Morange 2002; Nicoglou and Merlin 2017), I argue that Waddington's influence continues to inspire modern thinking about form and the structure and function of the genome.

5.2. From epigenesis to epigenetics

Waddington (1905–1975) has been described as "the last Renaissance biologist?" (Slack 2002), "a true twentieth-century polymath". And it was probably his diverse background and eclectic interests that contributed to his intellectual influence. Waddington trained at the University of Cambridge, England, where he studied natural sciences and specialized in geology. His graduate research was in paleontology, studying the structure of ammonites. This echoed his childhood interest in collecting fossils (Slack 2002; van Speybroeck 2002). I think we can speculate that these origins in geology and paleontology are at the heart of his life-long obsession with form. The word *epigenetic* was used in geology in the early 20th Century: an *epigenetic deposit* refers to a mineral deposit that formed later than the enclosing rocks (in contrast to a syngenetic deposit, which forms at the same time as the enclosing rocks). The mineral composition and chemical composition of epigenetic deposits differ from those of the enclosing rocks. Epigenetic deposits usually occur in the form of veins, lenses, stocks, and pipes that cut through the rocks. This training and interest in geology probably sowed the seeds for Waddington's thinking about the cumulative effects of time on form. To me this echoes the origins of Charles Darwin's own adventures in theoretical biology and the influence of geology on his thinking. It is well-documented that Darwin's most precious object that accompanied him on the voyage of the *Beagle* in 1831 was his copy of Lyell's *Principles of Geology*. He writes in his autobiography:

> The investigation of the geology of all the places visited was far more important, as reasoning comes here into play. On first examining a new district nothing can appear more hopeless that the chaos of rocks; but by recording the stratification and nature of the rocks and fossils at many points, always reasoning and predicting what will be found elsewhere, light soon begins to dawn on the district and the structure of the whole becomes more intelligible. […] I feel sure that it was this training which has enabled me to do whatever I have done in science (cited in Howard 1982, p. 3).

Just as Darwin's geological musings led him to evolution, Waddington abandoned paleontology to turn his interests towards embryology. Rather than the "chaos of rocks", Waddington was fascinated with how living organisms develop their complex forms (perhaps the "chaos of cells"). In the 1930s, he performed a series of experiments on chick embryos, inspired by the pioneering work of the German scientists Hans Spemann and Hilde Mangold, in search of the chemical nature of the inductive effect of the organizer (Slack 2002). These experiments strengthened his continued interest in the origins of "form" in complex biological systems. In the first chapter of his 1957 book *The Strategy of the Genes: A Discussion of Some Aspects of Theoretical Biology*, Waddington summarizes his thoughts.

> All living organisms, except possibly some of the very simplest, though even that exception is doubtful, are characterized by possessing a characteristic form or shape. This is not a mere static configuration, in which the organism happens to be produced and which it simply retains, as a stone or a mass of iron might do. The form or a living plant or animal is continuously kept in being in spite of the fact that material is passing all the time through the system. The form is not only, in many cases, a complex one, but the entity by which it is expressed is more comparable to a river than to a mass of solid rock. Finally, organic forms have a quality, difficult to express precisely in words, but rather definitely recognizable in practice, which is often referred to as "wholeness" or "integration". This is an almost aesthetic quality – a character of self-sufficiency and completeness (Waddington 1957).

A decisive visit to the *Drosophila* laboratory (the famous "*Fly Room*") of Thomas Hunt Morgan in the mid-1930s convinced Waddington of the interest of studying genes to understand developmental biology. In 1939, he wrote a students' handbook, entitled *Introduction to Modern Genetics,* in which he stressed the importance of linking genetics to the different branches of biology. In discussing developmental processes he stated that "the interaction of these constituents [of the fertilized egg] gives rise to new types of tissue and organ which were not present originally, and in so far development must be considered as 'epigenetic'" (Waddington 1939). It is noteworthy that here the word is used as an adjective, as it is used in geology, relating to the Aristotelian concept of "*epigenesis*" which refers to the development of animals or plants through the sequential interactions between constituents and the emergence of complex form over time (Nicoglou 2018). This is in contrast to *preformation* theories in which organisms are somehow fully developed in the egg or the sperm and embryonic development is simply an increase in size. A few years later, Waddington added an "*s*" and defined *epigenetics* as "the branch of biology which studies the causal interactions between genes and their products, which bring the phenotype into being". He insisted in the importance of this new field that would integrate genetics with embryology and developmental biology and explore the genotype-phenotype relationship. Waddington seems to have enjoyed the ambiguity about the switch from *epigenetic* to *epigenetics*. He wrote,

> This is derived from the Greek word *epigenesis*, which Aristotle used for the theory that development is brought about through a series of causal interactions between the various parts; it also reminds one that genetic factors are among the most important determinants of development. It is, however, not yet in common use (Waddington, 1956).

And it was around this time that he set up his laboratory in Edinburgh, embedded in the Scottish hillsides.

5.3. The evolution of the epigenetic landscape

Perhaps Waddington's most long-lasting legacy is the visual representation of the *epigenetic landscape*, the fruit of combining his interests in geology, genetics, embryology and his connection with artists (Baedke 2013). The first representation appeared on the cover of his book *Organisers and Genes* (Waddington 1940) and was a painting by his friend, the artist John Piper (see Figure 5.1a). Waddington wrote,

> Looking down the main valley toward the sea. As the river flows away into the mountains it passes a hanging valley, and then the two branch valleys, on its left bank. In the distance the sides of the valleys are steeper and more canyon like (Waddington 1940).

The idea was that developmental pathways were like the bifurcations of different streams as they descended the mountain towards the sea. This allowed Waddington to introduce the concept of *canalization* in which the depth of the valleys would affect the commitment to a given developmental pathway. Waddington was clearly satisfied with this visualization of his conceptual framework, commenting,

> I think it expresses, formally at least, some characteristics of development [e.g. canalization] which are not easy to grasp in any other way (Waddington 1940).

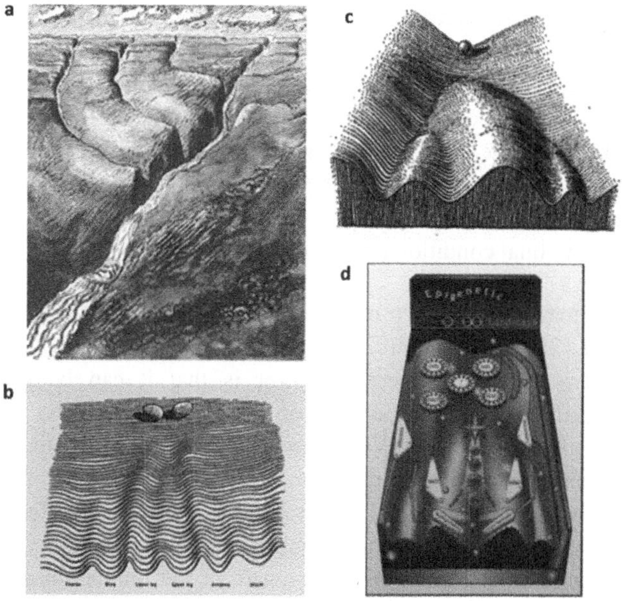

Figure 5.1. *Evolution of a landscape. Here are four visual representations of the "epigenetic landscape". (a) The earliest rendition appeared in* Organisers and Genes *(Waddington, 1940) and was a painting by the artist John Piper; (b) a later version placed two egg-shaped balls on the top of the mountain and different tissue identities at the bottom (Waddington, 1956); (c) the most well-known depiction appeared in* The Strategy of the Genes: A Discussion of Some Aspects of Theoretical Biology *(Waddington, 1957); (d) a modern landscape drawn as a pinball machine with multiple enzymes and effectors sending the ball in many directions drawn by Sue Ann Fung-Ho (Goldberg, et al., 2007) (used with permission from the author). For a color version of this figure, see www.iste.co.uk/chapouthier/life.zip*

Waddington was a prolific writer and used subsequent books to develop the idea further, both conceptually and visually (Baedke 2013; Nicoglou 2018). Later versions attempted to integrate genes, tissues and cellular differentiation states into the visual metaphor. For example, the 1956 version (see Figure 5.1b) has two egg-shaped balls on the top of the mountain, presumably ready to roll down the hill and be canalized into different tissular forms (Waddington 1956). A year later, he gave us a more refined version in his book *The Strategy of the Genes: A Discussion of Some Aspects of Theoretical Biology* (Waddington 1957). Now, there is a single "ball" at the top of the mountain, presumably the fertilized egg. In the book he explains:

> Consider a more or less flat, or rather undulating, surface which is titled so that points representing later states are lower than those representing earlier ones. Then if something, such as a ball, were placed in the surface it would run down towards some final end state at the bottom edge. [...] We have to picture the surface as grooved by valleys, each leading to one of the normal end states. The number of separate valleys must increase as we pass down from the initial towards the final condition (Waddington 1957).

This visual metaphor has been much discussed and is often cited (Baedke 2013; Goldberg *et al.* 2007; Nicoglou 2018; Slack 2002; Van Speybroeck 2002). It sometimes seems that its conceptual power is that it can be understood and interpreted in different ways, depending on the historical context and intellectual objectives. By tilting the plain, Waddington's landscape added the integration of time in the emergence of biological form. And the canalization into the valleys suggested that early developmental *choices* would commit to certain lineages and cell fates. The valleys at the bottom of the mountain may represent tissues, or cell states or differentiated cell types. He was aware of the limitations and ambiguities that the landscape presented, commenting that,

> It is worth emphasizing again that the epigenetic landscape is a representation in three dimensions of properties of the system which really involve many dimensions of phase-space. However, it does successfully incorporate some features of this system... (Waddington 1957).

He even adds,

> Although the epigenetic landscape only provides a rough and ready picture of the developing embryo, and cannot be interpreted rigorously, it has certain merits for those who, like myself, find it comforting to have some mental picture, however vague, for what they are trying to think about (Waddington 1957).

5.4. Modernizing the epigenetic landscape

Conrad Waddington died in 1975. Our understanding of gene regulation was relatively rudimentary and the genome sequencing was still a dream. Much has been written about how Waddingtonian epigenetics evolved into a modern understanding of the relationship between genotype and phenotype (Goldberg *et al.* 2007; Haig 2004; Morange 2002; Nicoglou and Merlin 2017). Over the years, the term "epigenetics" has taken on new meaning and new definitions, leading to a fair amount of confusion in the field. David Nanney's "*epigenetic control systems*" introduced the concept of cellular memory and stably-inherited patterns of gene expression (Nanney 1958). Subsequent work by Arthur Riggs and by Robin Holliday investigating the molecular mechanisms of inheritance defined a critical role for DNA methylation, leading to a mechanistic understanding of epigenetic phenomena (Holliday 2006; Riggs 1975). This led us to the contemporary, post-genomic definition as

> the study of any potentially stable and, ideally, heritable change in gene expression or cellular phenotype that occurs without changes in DNA sequence (Goldberg *et al.* 2007).

This shift of epigenetics, from the cellular to the molecular, has been accompanied by an "explosion" of discoveries linking epigenetic mechanisms to many aspects of biology. Epigenetics has been linked to paramutation in maize, position effect variegation in fruit flies, genomic imprinting of parental-specific loci in mammals and transgenerational epigenetic inheritance of stress or trauma. The rolling Scottish hills of Waddington's epigenetic landscape have been replaced by a *pinball* metaphor, in which the ball is bounced around by a combination of enzymes and effectors, writers and readers of epigenetic information (see Figure 5.1d) (Goldberg *et al.* 2007). The ball is cast about in a "chaos of genes"; there is a lot of noise and light, a lot of hype. The work of Jablonka and other philosophers pushes the concept of epigenetics further and further, with ambitions to link to contemporary evolutionary theory and behavioral inheritance (Jablonka and Lam 2005; Nicoglou and Merlin 2017). In an effort not to be restricted to molecular events in the nucleus or modifications of genomic chromatin, definitions with broader reach have been proposed; such as

> the study of various intracellular factors that have an effect on the stability of developmental processes though their action on genome potentialities (i.e. the genome susceptibility to be differentially expressed (Nicoglou and Merlin 2017).

What is clear is that we are not yet at the bottom of the Waddingtonian mountain; the "ball" continues its pathway downwards towards the valleys, perhaps at an ever increasing pace. But its final resting place is still of unknown destiny. There is still much to discover and epigenetic forms.

5.5. From epigenetic landscape to chromosome conformation

Despite the somewhat ambiguous and vague interpretations of the epigenetic landscape metaphor in some of Waddington's writing, he was also keen to speculate on the mechanisms underlying the developmental pathways that led to the emergence of form. The work of his contemporaries, such as the description of the structure of DNA by Watson and Crick (1953) or the operon model of gene regulation by Jacob and Monod (1961), obliged Waddington to integrate further the discoveries in molecular genetics with his "big-picture" vision of developmental biology. In his book *The Strategy of the Genes*, Waddington presents a prophetic vision for the underlying molecular mechanisms (Waddington 1957). Once again it is a rather "*rough and ready picture*", but it also has "*certain merits for those who, like myself, find it comforting to have some mental picture*". As Waddington himself explains,

> It is important to realize that the comparatively simple orderliness of the epigenetic landscape – its restricted number of valleys with their branching-points and characteristic contours – is a property of a higher order dependent on an underlying network of interactions which is vastly more complicated. [...] Since each gene must be regarded as a distinct chemical entity, the path of development as it is observed by the anatomist must be viewed as the resultant of all the very numerous processes in which these genes are involved in the cells concerned (Waddington 1957).

At this point, Waddington offers us another image (see Figure 5.2a) of the "underside" of the epigenetic landscape. Now, the "epigenetic surface is represented as a great sheet of canvas suspended at a considerable height above our heads and loping down towards the ground in the distance". Waddington draws a series of "*pegs and guy ropes*": "the pegs represent the genes," he suggests, "and the tensions of the guy ropes the chemical tendencies of the many genes; and if any gene mutates, altering the tension in a certain set of guy ropes, the result will not depend on that gene alone, but on its interactions with all the other guys". This visionary image of gene networks predicts what we refer to today as *systems biology*. Indeed, the conclusion of the Human Genome Project and the tools to manipulate the expression of the two dozen thousand human genes has led to an "*explosion*" of research on the "*pegs and guy ropes*" and how they impact the epigenetic landscape.

The study of epigenetics shifted from a concern with the emergence of form and the genotype-phenotype dilemma, to a veritable obsession with the molecular mechanisms that contribute to epigenetic processes. The last two decades have provided unprecedented insights into the "epigenetic tool-box", the molecular players that contribute to sculpting the landscape. These have been extensively reviewed elsewhere and are difficult to summarize in a limited space. The most well characterized is DNA methylation, the covalent addition of a CH3 to the 5th carbon atom on cytosine residues (Suzuki and Bird 2008). As mammalian DNA methylation is mostly in CpG dinucleotides, it offers an attractive opportunity to mark both strands of DNA with a covalent modification that can be transmitted to both daughter cells following DNA replication, thereby offering an insight into mechanisms of stable, mitotic inheritance of epigenetic information. Methylated DNA can be recognized by specific methyl-binding proteins and these in turn contribute to repression of gene expression, gene silencing and the formation of non-permissive heterochromatin. However, many of the other "epigenetic marks" are not on the DNA itself, but on the associated proteins that make up chromatin, the higher-order structure that forms when DNA wraps around histone proteins to form nucleosome structures. This organization allows compaction of the DNA molecule and offers a spatial organization that makes some portions of the genome more access ("euchromatin") than other closed portions ("heterochromatin"). Indeed, James Watson himself reflected that

> the major problem, I think, is chromatin [...] you can inherit something beyond the DNA sequence. That's where the real excitement of genetics is now (Watson 2003).

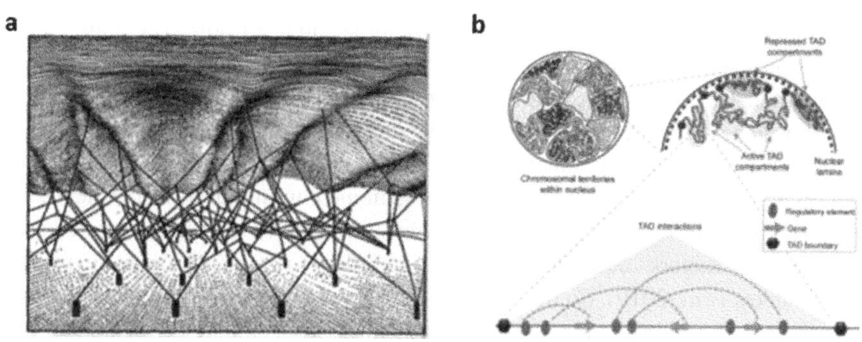

Figure 5.2. *Molecular forms under the landscape. (a) The underside of the epigenetic landscape showing the "pegs and guy ropes", the molecular mechanisms that link the action of genes to the form of the landscape (Waddington, 1957). (b) A representation of the 3D Genome showing how the genome is organized into Topologically Associated Domains (TADs) with functional consequence for gene regulation (from Matharu and Ahituv, 2015) (used with permission from the authors). For a color version of this figure, see www.iste.co.uk/chapouthier/life.zip*

The discovery that histones can be modified covalently in multiple ways (acetylation, methylation, phosphorylation etc.) at multiple positions led to the hypothesis that this information held the clues to understand genome regulation (Jenuwein and Allis 2001). This led to a new "explosion" – a global industry of mapping the modified histone marks across the genome and creating "epigenome" maps for human cells in normal and pathological states. Researchers believe that Waddington's landscape can now be understood in molecular terms – each position on the mountain represents a cellular state that can be described as a set of epigenome map features. The enzymes responsible for positioning these epigenetic modifications ("writers") and removing the modifications ("erasers") have been identified and characterized. And small-molecule inhibitors are emerging as the "epi-drugs" of a future epigenetic medicine. Indeed, we are beginning to master Waddington's "*pegs and guy ropes*" with a view to changing the form of the landscape to treat disease.

There is another aspect of "form" that I would like to discuss before concluding. Ever since Watson and Cricks' discovery of the double-helix structure, there has been a focus on the DNA molecule and the sequence of nucleotide building blocks from one end of the chromosome to the other. The second half of the 20th Century was characterized by a DNA-centric view of genetics, inspiring the unprecedented achievement of sequencing the entire human genome. But the post-genomic era has brought many challenges for understanding genotype-phenotype relationships. Exciting new tools and technological approaches are emerging that allow us to abandon this one-dimensional view of the genome and to explore its complex three-dimensional shape within the nucleus. This has become known as the "*3D Genome*" (Bonev and Cavallu 2016; Stadhouders *et al.* 2019). If Waddington's vision of a complex gene network of interrelated events is true, then we cannot investigate individual genes and their functions as if they existed in isolation, as independent loci on a long, inert string. It turns out that genome conformation is a critical factor of what might be called the *chromatin landscape* which regulatory proteins and transcription factors need to navigate in order to regulate gene expression (Stadhouders *et al.* 2019). So the "form" of the genome (in space and over time) will contribute to the cellular state, the transcriptional readout and the cellular phenotype. The development of an elegant technique called chromosome conformation capture (3C) enabled researchers to explore this chromosomal structure (Dekker *et al.* 2002). The eukaryotic nucleus contains *territories* for specific chromosomes and these territories are organized into chromosome compartments of "active" or "inactive" chromatin. Along the linear chromosome, genomic regions are organized into *topologically associating domains* (called TADs) that allow the formation of large extruding chromatin loops that regulate genome function (see Figure 5.2b). Thus, the shape of the genome in 3D space and the regulation of the forms of the 3D genome has a major impact on which genes are expressed at what time in which tissues. The folding of the 3D genome and the chromatin landscape appear to shape gene expression programs and affect cell-fate determination (Stadhouders *et al.* 2019). The TADs are often presented visually as triangles that resemble mountains emerging from a linear genome. Interestingly, unlike Waddington's mountains, these TAD images rise from the genome, a mountain of data constructed from the DNA at the bottom to the peaks of interaction and regulatory contacts at the top. But they emphasize the complex regulatory networks (the "*pegs and guys ropes*" perhaps) that result from and contribute to genome function and the form of the epigenetic landscape.

5.6. Conclusion: from form to function

From the very beginning, *epigenetics* was linked to discussions of form. By bringing together genetics and developmental biology, Waddington reintroduced thinking about *epigenesis*, about the emergence of form over time through the interactions between constituent parts. This led to the evolution of a *landscape* metaphor to describe the cell-fate determination over developmental time. The challenge to understand the relationship between genotype and phenotype (and in a larger context between Nature and Nurture) drove him to describe a conceptual framework. By adding an "s" and switching from an adjective to a noun, he created a whole new field of biology and inspired generations of researchers. Although, the word *epigenetics* continues to be re-defined, the desire to understand how form emerges, and how form impacts function, remains a motivating challenge. Indeed, the new discoveries of the 3D genome and nuclear compartmentalization has fueled debate about whether conformational TADs are functionally relevant and whether they are the drivers or the outcomes of differential gene expression. In some ways, the "*chaos of the genes*" continues to intrigue us and we are perhaps further than we think from understanding the rules with which to master the "*pegs and guy ropes*" in any meaningful (or clinically pertinent) way. The "explosion" of sophisticated technological approaches to address these issues leaves me somewhat optimistic about our future work relating form to function. But perhaps I can offer the final word to Waddington himself. He concludes *The Strategy of the Genes* with the following thoughts:

> Where there is no immediate prospect of experimental verification of a model, it may, nevertheless, not be without value. […] The newer methods and techniques which today are so successfully applied in many fields of biology can become more than better tools. In a sense they can create new problems, and allow us to ask more searching questions to which our curiosity demands an answer (Waddington, 1957).

5.7. Acknowledgments

Work in the Weitzman laboratory was supported by the LabEx "Who Am I?" #ANR-11-LABX-0071 and the Université de Paris IdEx #ANR-18-IDEX-0001 funded by the French Government through its "Investments for the Future" program. JBW is a senior member of the Institut Universitaire de France (IUF).

5.8. References

Baedke, J. (2013). The epigenetic landscape in the course of time: Conrad Hal Waddington's methodological impact on the life sciences. *Stud. Hist. Philos. Biol. Biomed. Sci.*, 44, 756–73.

Bonev, B., Cavalli, G. (2016). Organization and function of the 3D genome. *Nat. Rev. Genet.*, 17, 772.

Dekker, J., Rippe, K., Dekker, M., Kleckner, N. (2002). Capturing chromosome conformation. *Science*, 295, 1306–11.

Goldberg, A.D., Allis, C.D., Bernstein, E. (2007). Epigenetics: a landscape takes shape. *Cell*, 128, 635–638.

Haig, D. (2004). The (dual) origin of epigenetics. *Cold Spring Harb. Symp. Quant. Biol.*, 69, 67–70.

Holliday, R. (2006). Epigenetics: a historical overview. *Epigenetics*, 1, 76–80.

Howard, J. (1982). *Darwin: A Very Short Introduction*. Oxford University Press, Oxford.

Jablonka, E., Lamb, M.J. (2005). *Evolution in Four Dimensions: Genetic, Epigenetic, Behavioral, and Symbolic Variation in the History of Life*. MIT Press, Cambridge.

Jacob, F., Monod, J. (1961). Genetic regulatory mechanisms in the synthesis of proteins. *J. Mol. Biol.*, 3, 318–356.

Jenuwein, T., Allis, C.D. (2001). Translating the histone code. *Science*, 293, 1074–1080.

Matharu, N., Ahituv, N. (2015). Minor loops in major folds: Enhancer–promoter looping, chromatin restructuring, and their association with transcriptional regulation and disease. *PLoS Genet.*, 11.

Morange, M. (2002). The relations between genetics and epigenetics: A historical point of view. *Annals of the New York Academy of Sciences*, 981, 50–60.

Nanney, D.L. (1958). Epigenetic control systems. *Proc. Natl. Acad. Sci. USA*, 44, 712–217.

Nicoglou, A. (2018). Waddington's epigenetics or the pictorial meetings of development and genetics. *Hist. Philos. Life Sci.*, 40, 61.

Nicoglou, A., Merlin, F. (2017). Epigenetics: A way to bridge the gap between biological fields. *Stud. Hist. Philos. Biol. Biomed. Sci.*, 66, 73–82.

Riggs, A.D. (1975). X inactivation, differentiation, and DNA methylation. *Cytogenet. Cell. Genet.*, 14, 9–25.

Slack, J.M. (2002). Conrad Hal Waddington: the last Renaissance biologist? *Nature Reviews Genetics*, 3, 889–95.

Stadhouders, R., Filion, G.J., Graf, T. (2019). Transcription factors and 3D genome conformation in cell-fate decisions. *Nature*, 569, 345–354.

Suzuki, M.M., Bird, A. (2008). DNA methylation landscapes: provocative insights from epigenomics. *Nat. Rev. Genet.*, 9, 465–76.

Van Speybroeck, L. (2002). From epigenesis to epigenetics: the case of C.H. Waddington. *Ann. NY Acad. Sci.*, 981, 61-81.

Waddington, C.H. (1939). *An Introduction to Modern Genetics*. Allen and Unwin, London.

Waddington, C.H. (1940). *Organisers and Genes*. Cambridge University Press, Cambridge.

Waddington, C.H. (1942). The epigenotype. *Endeavour*, 1, 18–20.

Waddington, C.H. (1956). *Principles of Embryology*. Allen and Unwin, London.

Waddington, C.H. (1957). *The Strategy of the Genes: A Discussion of Some Aspects of Theoretical Biology*. Allen and Unwin, London.

Watson, J.D. (2003). Celebrating the genetic jubilee: a conversation with James D. Watson. Interviewed by John Rennie. *Sci. Am.*, 288, 66–9.

Watson, J.D., Crick, F.N. (1953). Molecular structure of nucleic acids; a structure for deoxyribose nucleic acid. *Nature*, 171, 737–8.

6

The Many Shapes of Microbial Detection of Kin and Kind

Guillermo PAZ-Y-MIÑO-C and Avelina ESPINOSA
Roger Williams University, Bristol, United States

Single-celled organisms, or microbes, belong to two groups: those with a circular chromosome, the prokaryotes, which include bacteria (e.g. *Escherichia coli*, a symbiont in the human gut) and archaea (like *Sulfolobus solfataricus*, frequent in hot springs), and those with elongated, telomeric chromosomes inside a nucleus, the unicellular eukaryotes or protists (e.g. *Amoeba proteus*, common in aquatic environments). Ancestors of microbes lived on Earth for 3–4 billion years; their current descendants occupy all ecological niches, from free-living (most) to pathogenic (Adl *et al.* 2019; Parks *et al.* 2018; Paz-y-Miño-C and Espinosa 2018).

Bacteria, archaea and protists are anatomically/functionally equipped to detect close genetic relatives (clone or kin, which coefficient of genetic relatedness $r = 1.0$ or $r \approx 1.0$), as well as distantly related conspecifics (different clones/taxa, where $r < 1.0$ or $r \ll 1.0$ or $r \approx 0.0$). Why? In this chapter, we discuss the adaptive value of a microbe's ability to discriminate/recognize kin from non-kin. Because the theoretical framework of this field (= kin discrimination/recognition) was initially developed for animals – and some plants (Tang-Martínez 2001; Penn and Frommen 2010; Dudley *et al.* 2013), we shall first explain its foundations and later examine unicells' examples. Our emphasis will be on "shapes" (cell aggregations for cooperation with other unicells closely related in their genomes = kin) and genetics (the replicators for discrimination/recognition of kin or of "cells of the same kind", i.e. those similar only in the functions of some genes, like immunities to toxins, but distinctive in their genomes (Paz-y-Miño-C and Espinosa 2018)). We will highlight

coordinated by Georges CHAPOUTHIER and Marie-Christine MAUREL. © ISTE Ltd 2020.

the relevance of kin detection for altruistic cooperation (including among pathogens) or for the identification of "cheater cells" (= the beneficiaries of others' selfless acts, but that do not contribute to the collective well-being), and for the formation of temporary or permanent cell alliances (e.g. predatory "swarms", dispersing "slugs", substrate-anchoring "mounds", biofilms, fruiting bodies for spore production), which are informative for understanding the evolutionary origins of multicellularity.

6.1. From Darwin's social-insects-puzzle to microbes

Why do organisms preferentially cooperate with close genetic relatives rather than with strangers? In *The Origin of Species* (1859), Charles Darwin speculated about the "puzzle of the sterile social insects", in which female workers at a nest dedicated their lives to the persistence of a colony by assisting a fertile queen to reproduce with the available males (Paz-y-Miño-C and Espinosa 2018). Darwin implied that in such cases of apparent workers' sacrifice for the good of the community, "selection may be applied to the family". But, in the 1800s, he could not offer a heredity-based explanation for "selection [being] applied to [kin]". It was not until Ronald A. Fisher (1930) and John B. S. Haldane (1932, 1955) wrestled with the mathematics of altruism that the anecdotal expression "I would lay down my life for two brothers or eight cousins" became the legacy of their work (Paz-y-Miño-C and Espinosa 2018). William D. Hamilton (1963, 1964) and John Maynard-Smith (1964) later reasoned that the ability to discriminate between close and distant genetic relatives could be directly linked to survival and reproductive success, and, ultimately, to kin selection (Maynard-Smith 1964, 1977). Close relatives would engage in altruistic cooperation to pass on the shared genes and minimize kin competition. According to Hamilton, the fitness benefits (b) of helping another individual correlate with the coefficient of genetic relatedness (r) and are dependent on the costs (c) of helping. If the benefits (rb) surpass the costs (c), then altruistic cooperation could evolve ($rb - c > 0$). Hamilton's model suggested that prior to engaging in cooperation among relatives, organisms would have to detect kin and do it via a gene or genes (a mechanism later called "recognition alleles"; Paz-y-Miño-C and Espinosa 2016, 2018). Hamilton's "super gene" model referred to pleiotropic loci capable of: 1) influencing the expression of a trait for recognition, which would allow 2) carriers of the trait to recognize similar traits in others, and 3) induce the carriers to behave altruistically only toward other carriers of the gene. Richard Dawkins (1976) called this phenomenon the "green-beard effect" (= green-beard genes) in which any carrier of such "tag" would spot it in conspecifics.

Thus, theory and experiments to elucidate the evolution of kin discrimination/ recognition initially came from studies with multicellular eukaryotes, particularly social insects (ants, bees), and later vertebrates and plants (above). The concepts of natural selection (a mechanism itself) in the 1850s (Darwin's work), inclusive fitness (Hamilton's) and kin selection (Maynard-Smith's), the latter two in the 1960s–1970s, became the foundation of kin discrimination/recognition at the mechanistic level (= how organisms managed to tell apart kin versus non-kin via, as just mentioned, recognition alleles, or, as it was later discussed in the 1980s, via early-life learning to identify kin, as inferred from observations in mammals, including humans; Paz-y-Miño-C and Espinosa 2018). During the 2000s, studies with microbes became novel and focused on the genetics of cell-cell discrimination/recognition (West *et al.* 2006; Kraemer and Velicer 2011; Celiker and Gore 2013; Mitri and Foster 2013; Espinosa and Paz-y-Miño-C 2014a, 2019; Espinosa *et al.* 2016; Paz-y-Miño-C and Espinosa 2016, 2018). This last approach nurtured the previous knowledge (1960s onwards), but it soon expanded toward the study of the evolution of multicellularity and cheating. Presently, the field is centered on the genetics, evolution, behavior and health aspects of the cell-to-cell mechanisms of communication, the factors leading to facultative or permanent aggregations, as well as on mathematical modeling to exploring the complexity of organismal systems built upon billions of clonal, cooperating cells (e.g. a blue whale), and in which microbes are central to the simulations (Paz-y-Miño-C and Espinosa 2018).

6.2. Handshakes of kinship or "kindship" in bacteria

The rod-shaped bacterium *Myxococcus xanthus* lives in the soil and forms aggregations that forage on organic matter or actively pursue prey in swarms. Cells (myxobacteria) proliferate vegetatively via bipartition and engage in foraging journeys (millions of individuals) that include facultative "hunting" of other cells (Figure 6.1; Paz-y-Miño-C and Espinosa 2018). Large swarms separate into smaller ones and scattered, peripheral rods establish new aggregations that, if resources suffice, form swarms of their own. During food scarcity or drought, the cells survive by morphing into spores (tolerant to starvation, heat, desiccation), but prior to that, swarms go through the stage(s) of mound-and-fruiting-body formation (mounds anchor onto a substrate, while the fruiting body grows on top and produces spores). This transformation requires cell-autolysis or programmed cell death (PCD) at the base of the fruiting body, otherwise the myxospores cannot develop (they need optimal above-substrate microenvironment). The spores germinate upon improved conditions (nutrients, temperature, humidity), new rods emerge from the spore-encasements and rod proliferation continues (Paz-y-Miño-C and Espinosa 2018).

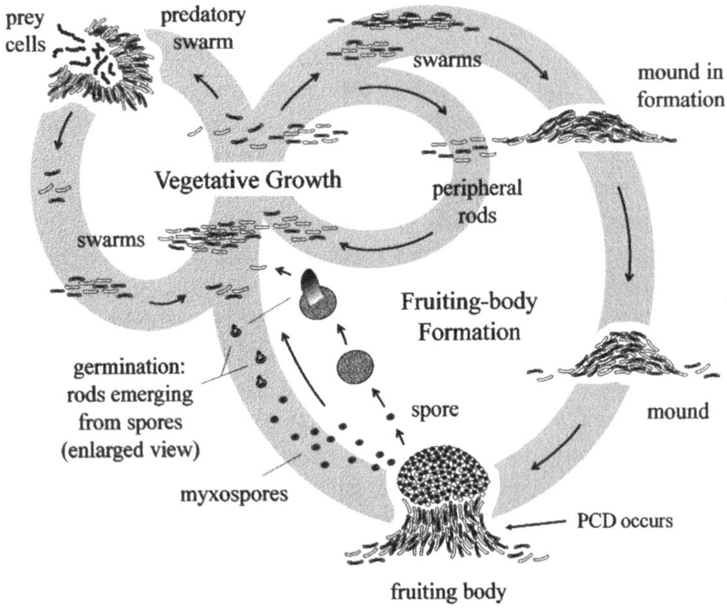

Figure 6.1. *Myxococcus xanthus lifecycle (source: Paz-y-Miño-C and Espinosa 2018)*

Myxococcus swarms and fruiting bodies belong to populations of highly related bacteria (clones or "self-like" members). Genetically distant competitors are excluded from these aggregations and killed or inhibited by toxins excreted by the prevalent strain (Wall 2016). Such toxins (bacteriocins) bind to cell membrane receptors of rivals. Clone members are immune to them.

Strain discrimination/recognition is central to myxobacteria's ability to cooperate with highly-related individuals, minimize non-adaptive cooperation with strangers (non-related or distantly related), and avoid cheaters (Wall 2016; Cao *et al.* 2015; Paz-y-Miño-C and Espinosa 2018). Scientists have identified a polymorphic gene (*traA*), whose functions are consistent with those of a green-beard allele (above). *traA* encodes for outer membrane (OM) lipoproteins (LP) associated with cell-cell mutual recognition (via the cell-membrane receptor TraA). Specificity for discrimination/recognition is given by variable sequences in the *traA* alleles. *traA* facilitates swarming among self-like partners (as per "motility proteins" exchange) and fruiting-body formation, and, when genetically compatible cells encounter one another, *traA* mediates the exchange of OM-LP between cells carrying specific lipoproteins in the OM and transient non-carriers, or non-carrier mutants. OM-LP exchange allows cells to give/receive temporary immunity to their own strain's bacteriocins (Pathak *et al.* 2013; Cao *et al.* 2015; Paz-y-Miño-C and Espinosa 2018).

Myxococcus cells can fuse their outer membranes and swap OM-LP. When a cell lacks, for example, a "gliding" motility protein on its OM, it can procure it from a compatible donor, likely a close relative (Figure 6.2; Paz-y-Miño-C and Espinosa 2018). The transaction is mutually beneficial (e.g. motility proteins, immunity factors, repair of damaged OM), but, if altruistic, the cost to the donor, at the expense of giving away a valuable cell product, should translate into genetic gain according to Hamilton's $_rb - c > 0$. Bacteria equipped with motility proteins participate in swarming (foraging, predation, dispersal), as well as mound and fruiting-body formation (Pathak *et al.* 2013; Cao *et al.* 2015; Wall 2016). These behaviors entail risks (= PCD) and only cooperation with the right partners increases survival/reproduction (perpetuation of shared genes). Individual cells cannot persist long-term, and always-cheaters cannot prevail (Paz-y-Miño-C and Espinosa 2018).

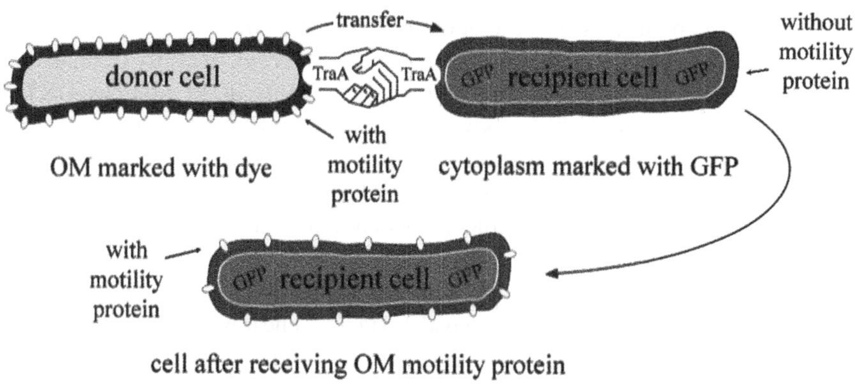

Figure 6.2. *Transfer of an outer membrane (OM) motility protein from a donor to a recipient Myxococcus bacteria (source: Paz-y-Miño-C and Espinosa 2018)*

As mentioned earlier, TraA acts as a cell surface receptor. Note in Figure 6.2 that donor and recipient cells are equipped with TraA compatible proteins on their OM lipoproteins (TraA–TraA "shaking hands") that allow discrimination/recognition among matching partners prior to the OM motility-protein exchange-transfer. In actual experiments, scientists marked with dye the motility proteins of donor cells and with green fluorescent protein (GFP) the cytoplasm of recipient cells. OM-LP transfer was inferred when the GFP-marked cells acquired the outer-membrane dye coloration of the donors (Pathak *et al.* 2013; Paz-y-Miño-C and Espinosa 2018).

To explore the taxa-, clone- and kin-discrimination/recognition role of *traA*, researchers replaced the *traA* allele of the laboratory strain *M. xanthus* DK1622 (*traA*$^{\mathrm{DK1622}}$) with that of a *M. fulvus* strain (*traA*$^{\textit{M. fulvus}}$). In preliminary tests, these two clones did not exchange-transfer OM-LP (Pathak *et al.* 2013). When the laboratory strain DK1622 expressed the borrowed *M. fulvus* allele, as DK1622 *traA*$^{\textit{M. fulvus}}$, the exchange-transfer of OM-LP between *M. xanthus* DK1622 and *M. fulvus* did occur. The resulting bioengineered bacteria behaved as "cells of the same kind" or "kindship" (= similar in the functions of *traA*, but distinctive in the rest of their genomes; Paz-y-Miño-C and Espinosa 2018). Comparable results were obtained when the *traA* allele of *M. xanthus* DK1622 (*traA*$^{\mathrm{DK1622}}$) was replaced with *traA* genes coming from three other *M. xanthus* donor strains (*traA*$^{\mathrm{DK816}}$, *traA*^{Pali}, *traA*$^{\mathrm{A96}}$ (Pathak *et al.* 2013)). In consequence, *traA* replacements did change discrimination/recognition partner specificity between somewhat distantly related genetic relatives: the researchers observed a "handshake of kindship" between cells in which a meaningful trait in common for the interaction was their compatible *traA*.

Additional experiments on swapping *traA* alleles among swarming strains led investigators to suspect that *traA* was linked to *Myxococcus'* motility (Pathak *et al.* 2013). The researchers noticed that non-motile mutants of *M. xanthus* DK1622 inhibited swarming and fruiting-body formation of motile strains (DK801, A66, A88). The inhibition was clone specific as per overall genetic proximity between strains (Pathak *et al.* 2013; Paz-y-Miño-C and Espinosa 2018). This suggested that wild populations of non-motile/motile strains co-depended to either cooperate with "same" or opportunistically supplant rivals ("different") when the environmental conditions were right (Paz-y-Miño-C and Espinosa 2018). Remember that mound and fruiting-body formation are costly (PCD, cheating). Thus, via clone discrimination/recognition, *traA* mediated the interactions before self-like bacteria committed fully, altruistically and adaptively to aggregations and motility, which, ultimately, led to investing (as per $_rb - c > 0$) in survival and reproduction.

Since *traA* mediates OM-LP exchange-transfer between close strains, can it also mediate sharing of immunity? This question is pertinent because, as we discussed above, *Myxococcus* can secrete bacteriocins that kill unrelated cells and/or inhibit their reproduction. In addition, it is known to biologists that the cell outer membrane participates in cell immunity. During OM-LP exchange-transfer experiments (comparable to Figure 6.2), investigators observed that the *M. fulvus* strain killed the laboratory strain *M. xanthus* DK1622 after six hours of mixed incubation. When the researchers inserted the *traA*$^{\textit{M. fulvus}}$ allele into the *M. xanthus* DK1622 strain (replacing *traA*$^{\mathrm{DK1622}}$ with *traA*$^{\textit{M. fulvus}}$), the laboratory strain acquired partial immunity to the *M. fulvus* toxins and increased survival (Pathak *et al.* 2013; Paz-y-Miño-C and Espinosa 2018). The investigators also noticed that bacterial strains genetically similar to one another (e.g. DK1622 and DK801) did not kill each other in mixed cultures. But, when the *traA* allele was knocked out in DK1622 (Δ*traA*^{DK1622}) and both clones grown together (Δ*traA*^{DK1622} and *traA*^{DK801}), survival of the knocked-out clone Δ*traA*^{DK1622} decreased (Pathak *et al.* 2013; Paz-y-Miño-C and Espinosa 2018). These findings demonstrated that myxobacteria could kill or inhibit reproduction of genetically distant strains (*M. fulvus* vs *M. xanthus* DK1622) and that, by contrast, close genetic relatives were immunotolerant to each other's toxins (*M. xanthus* DK1622 vs DK801). Because immunity to toxins could be transiently acquired or improved via exchange-transfer of OM-LP between compatible cells (= TraA–TraA cell-surface receptors), *traA* acted as facilitator of the process (Pathak *et al.* 2013; Paz-y-Miño-C and Espinosa 2018).

In conclusion, *Myxococcus* bacteria can tell apart same (self/self-like) from different (non-self/non-self-like); *traA* plays a crucial role in cell discrimination/recognition (Paz-y-Miño-C and Espinosa 2018). As a green-beard, *traA* influences the expression of a trait for recognition (= TraA cell-surface receptor). This gene allows carriers of TraA to detect the same trait in others. *traA* induces the carriers of TraA to behave altruistically only toward other carriers of compatible TraA (OM-LP exchange-transfer). *traA* is polymorphic. Exchange-transfer of OM-LP between donor and recipient cells is costly (= motility proteins, immunity factors); thus, altruistic cooperation allows closely related partners (kin/kin-like) to survive and reproduce under environmental stress, and swarm for mound-and-fruiting-body formation and spore production, stages in which mortality is high. *traA* is also involved in adversary interactions (Paz-y-Miño-C and Espinosa 2018).

6.3. The ameba world of clone discrimination/recognition

In taxa- and clone-discrimination/recognition experiments conducted in our laboratory, we demonstrated that diverse species of the protist *Entamoeba* discriminate self from different (Espinosa *et al.* 2016). We first showed that seven *Entamoeba* cell-lines could be characterized by us – human observers – morphometrically as per the amebas' length, width, and cell-surface area (microscope observations): *E. histolytica* HM-1:IMSS, *E. dispar*, *E. moshkovskii* Laredo, *E. moshkovskii* cf. Snake, *E. terrapinae*, *E. invadens* IP-1 and *E. invadens* VK-1:NS. Of the three measurements we examined, cell length was the most informative for clone-morphometric distinction; length allowed us to differentiate among most clones. Cell width or soma surface area were less informative than length, but each allowed us to roughly group amebas of variable widths or soma areas. When we combined length and width, we could tell each clone apart with 95% confidence. Thus, morphometrics alone (single or combinations of variables) allowed us to discriminate among all cell lines.

Do ameba clones themselves/discriminate/recognize one another? Yes, and they do it among close or distantly related taxa (Figure 6.3). Realize first that our seven cell-lines are representative of major lineages within the genus *Entamoeba* (Espinosa and Paz-y-Miño-C 2019). Some, like *E. invadens* IP-1 and *E. invadens* VK-1:NS, differ only in a single nucleotide in the small-subunit-ribosomal-RNA sequence (Espinosa and Paz-y-Miño-C 2012, 2014a, 2014b), yet they discriminate one another as much as any other pair of phylogenetically more distant taxa (i.e. *Ei*-IP-1 and *Ei*-VK-1:NS are likely distinctive species, but only the *Ei*-IP-1 genome is known (Espinosa and Paz-y-Miño-C 2019)).

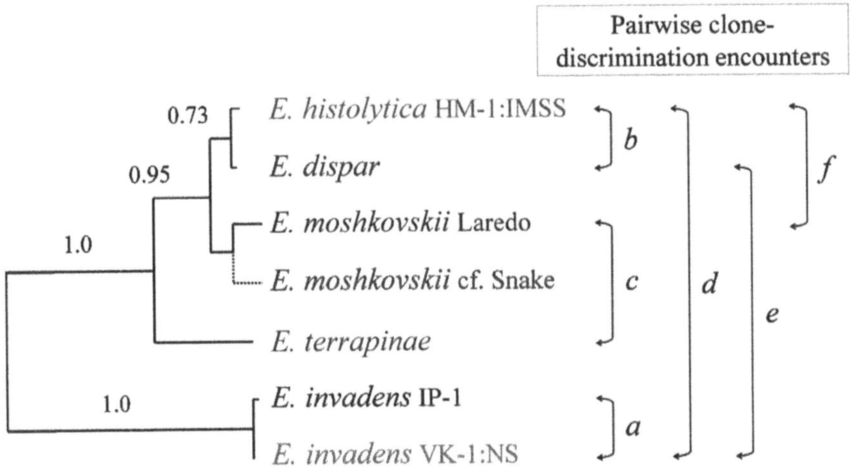

Figure 6.3. *Phylogeny based on ssrRNA sequences of Entamoeba clones. The placement of E. moshkovskii, see Snake (dashed-line) is tentative (=ssrRNA sequence in review). Numbers indicate statistical branch support, Approximate Likelihood Ratio Test (Espinosa and Paz-y-Miño-C 2019)*

As marked *a–f* in the connecting arrows in Figure 6.3, all cell lines segregate into single-clone clusters when grown in pairs in the laboratory (Espinosa and Paz-y-Miño-C 2019): most closely related sister taxa *E. invadens* IP-1 vs. *E. invadens* VK-1:NS (*a* in Figure 6.3); more distantly related sister taxa *E. histolytica* HM-1:IMSS versus *E. dispar* (*b*); between a representative of a known multi-species clade (e.g.

E. moshkovskii Laredo) versus the most distant member of that clade (*E. terrapinae*, *c*); as well as between members of two distinct clades (*E. histolytica* HM-1:IMSS versus *E. invadens* VK-1:NS, *d*; or *E. dispar* versus *E. invadens* VK-1:NS, *e*). We have obtained comparable results with *E. histolytica* HM-1:IMSS versus *E. moshkovskii* Laredo (*f*).

Figure 6.4 shows fluorescent micrographs of the exemplar co-growth of *E. histolytica* HM-1:IMSS versus *E. dispar* (for images of other pairs see Espinosa and Paz-y-Miño-C 2012, 2014a,b, 2019; Espinosa *et al.* 2016; Paz-y-Miño-C and Espinosa 2018). Figure 6.4a corresponds to *E. histolytica* HM-1:IMSS labeled green versus *E. dispar* labeled red; each clone aggregated separately. Figure 6.4b shows the reverse-color labeling of *E. histolytica* HM-1:IMSS (red) versus *E. dispar* (green); the clones also aggregated in distinct clusters. Figure 6.4c corresponds to *E. histolytica* HM-1:IMSS labeled with both green/red dyes; the cells mixed and looked yellow. Figure 6.4d shows *E. dispar* labeled with both green/red dyes (similar to c).

Figure 6.4. *Clone-aggregation preference in mixed cultures of E. histolytica HM-1 IMSS and E. dispar (a,b), or single-cell-line cultures (c,d). Images taken at 36-h scale bar = 100-μm, 10x magnification (source: Espinosa and Paz-y-Miño-C 2019). For a color version of this figure, see www.iste.co.uk/chapouthier/life.zip*

Thus, the mix-cell-line-cultures between phylogenetically close or distant lineages demonstrate that *Entamoeba* spp. are capable of taxa- and clone-discrimination/recognition. Equal abilities have been reported in free-living social amebas in the genera *Dictyostelium* and *Polysphondylium* (Queller *et al.* 2003; Ostrowski *et al.* 2008; Hirose *et al.* 2011; Kaushik *et al.* 2006; Mehdiabadi *et al.* 2006; Smith *et al.* 2016; Noh *et al.* 2018). The cell-lines in our laboratory correspond to free living, commensal and/or pathogenic (Espinosa and Paz-y-Miño-C 2019).

During the laboratory pair-encounters between clones, we selected another exemplar (*E. invadens* IP-1 vs *E. invadens* VK-1:NS; Figure 6.3) to chemically identify putative cell-signals secreted by the amebas during the co-growth (Espinosa *et al.* 2016; Espinosa and Paz-y-Miño-C 2019; Paz-y-Miño-C and Espinosa 2018). We found six proteins common to both cell lines: RasGap/Ankyrin, coronin-WD40, actin, protein kinases, heat shock 70, and ubiquitin. Some of their functions in *Entamoeba* spp. included cell proliferation, cell adhesion, cell movement (pseudopod formation assisted by actin polymerization), migration, and stress-induced encystation (all relevant in the context of cell-cell discrimination/recognition for aggregation (Espinosa and Paz-y-Miño-C 2019)).

In conclusion, *Entamoeba* clones of diverse taxa can be characterized – by human observers – morphometrically with 95% confidence. Amebas themselves can discriminate/recognize between same (clone) and different (other clones) and they do it by secreting chemical cues that induce amebas to aggregate preferentially with clone members. The differential behavioral treatment of clone- versus non-clone members, and between phylogenetically close (*E. invadens* IP-1 and *E. invadens* VK-1:NS) or distant taxa (*E. terrapinae* vs. *E. moshkovskii* Laredo), demonstrates that the capacity to discriminate between same and different in these unicellular eukaryotes is an ancient trait (because amebas, as protists, are ancient; Espinosa *et al.* 2016; Espinosa and Paz-y-Miño-C 2019; Paz-y-Miño-C and Espinosa 2018). We still need to determine the gene(s) involved in this ability, and if amebas can discriminate/recognize among strains of gradually-decreasing genetic relatedness (r), which would be an indication of kin-discrimination/recognition.

6.4. Social microbes and multicellularity

Biologists think that multicellularity originated – multiple times – in clonality (Fisher *et al.* 2013), via recruitment of closely related cells capable of recognizing one another, and which lived in proximity and benefited from specialized division of labor (Ruiz-Trillo and Nedelcu 2015; Brunet and King 2017; Paz-y-Miño-C and Espinosa 2018). Such specialization included the restriction of soma-reproduction to the gametes. Because sexually reproducing multicellular organisms (e.g. fruit flies, mice, humans, blue whales) experience a "generational single-celled bottleneck" (zygote) that forces its descendants to be clonal (r = 1.0), the evolution of multicellularity has been explained under inclusive fitness theory (Paz-y-Miño-C and Espinosa 2018). Research with microbes, however, has led investigators to propose alternative scenarios (e.g. tolerance to cheating, predation) under which multicellularity could have evolved (Ratcliff *et al.* 2012; Queller and Strassmann 2013; Tarnita *et al.* 2015; van Gestel and Tarnita 2017; Herron *et al.* 2019).

In Figure 6.5, we sketch some patterns of prokaryotic multicellularity (after Paz-y-Miño-C and Espinosa 2018), starting with the soil bacterium *Bacillus subtilis* (Figure 6.5a), which forms biofilmic clonal assemblages with cell specializations among participants (i.e. cells for anchoring on a substrate, matrix-secreting cells, reproductive layers of cells including the spores they produce, and flagellated dispersers). Biofilms get fragmented due to external disruptions or via dispersal of motile cells, which swarm, settle on new environments, and form biofilms again. In *Anabaena* (a planktonic cyanobacterium), individual cells divide and remain attached by proteins (septosomes), forming chains of cells (Wilk *et al.* 2011). In other bacterial species (e.g. the subaerial *Fischerella*), cell divisions generate chain-clusters with branching patterns (Figure 6.5b; Schirrmeister *et al.* 2011). Fragments of these structures separate, disperse and act as propagules for further proliferation.

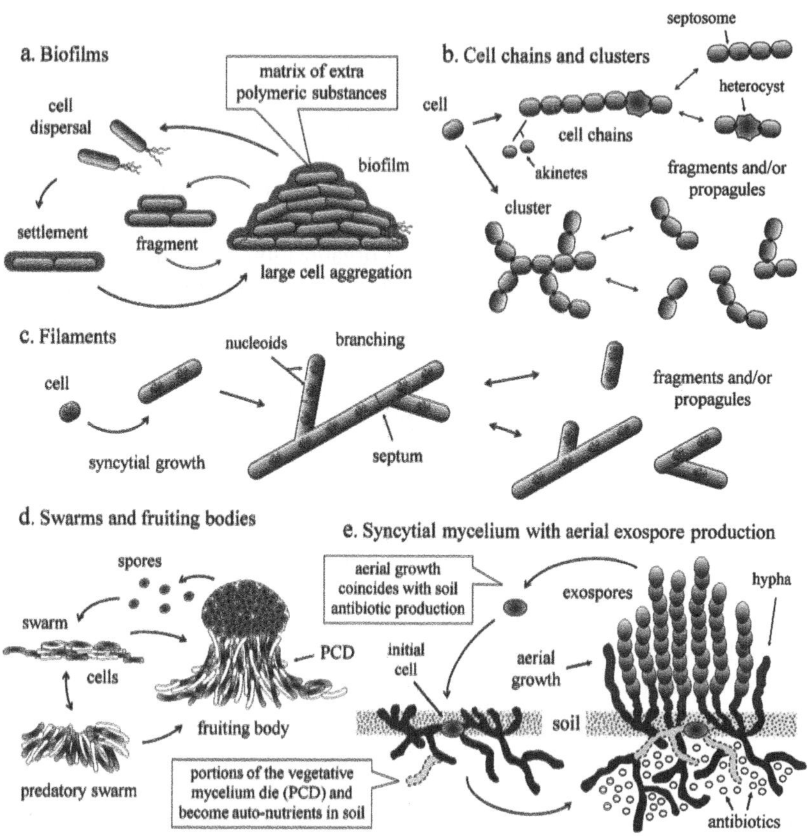

Figure 6.5. *Bacterial multicell-aggregations (source: Paz-y-Miño-C and Espinosa 2018)*

Syncytial growth (Figure 6.5c) relies on cytoplasmic expansion and generation of multiple clonal "nucleoids" (each containing bundled DNA originated in "bacterial-chromosome" divisions) inside a propagating tube (filament), which can branch off. Plasma membrane crosswalls (septa) form along the tube, allowing tube-fragments to separate and disperse as propagules; some fragments/propagules result from PCD of inter-septa protoplasmic pieces. Multicellular swarms and fruiting bodies in *M. xanthus* are another example of clonal aggregations with specialized division of labor (Figure 6.5d, also Figure 6.1). *Streptomyces coelicolor*'s lifecycle is also illustrative of cell cooperation/specialization (Figure 6.5e). This bacterium propagates syncytially via a vegetative mycelium, started by a cell that germinates in the soil, grows a network of protoplasmic tubes with multiple nucleoids, and extends aerial branches (hyphae) for spore production (exospores). The growth of the aerial branching co-occurs with PCD of the mycelium network, which turns into nutrients that are exploited by the above-surface hyphae. These events coincide with bacteriocin and antibiotic secretions to which the mycelium is immune; thus, securing that the "altruistic sacrifice" by the dying branches, protoplasmic tubes and nucleoids benefit kin (Claessen *et al.* 2014; Paz-y-Miño-C and Espinosa 2018).

How did multicellularity, as we know it today, get started? The gradual transition toward obligate multicellularity entailed the advantages of complex cooperation, large somatic size, and the hierarchical organization of cells that assembled themselves into groups, which, in turn, assembled into multicellular organisms (Michod and Roze 2001; van Gestel and Tarnita 2017; Paz-y-Miño-C and Espinosa 2018). Under this view, the ecology and heritable development of ancestral solitary forms (of differential aggregation potential) were crucial for the evolution of more permanent multicellular grouping (e.g. facultative multicellularity in the social ameba *Dictyostelium*, below, or permanent/obligate in humans). This was likely the case because the reproduction of current multicellular taxa is bound to stages within a lifecycle. Such stages can be represented by the sequence: adult multicellular individual → gamete → zygote or equivalent (spore or propagule; strict unicellular stage via population bottleneck) → embryo (clonal cell propagation) → adult.

Equivalents to the sequence above can be identified in the lifecycle of *Dictyostelium discoideum* depicted in Figure 6.6. Upon environmental stress, like starvation, thousands of soil free-living amebas aggregate in "mounds", which turn into "slugs" that move synchronously; slugs anchor on substrates to form fruiting bodies by allocating dying cells to stalks that support "sori", inside of which cells differentiate into spores; once released, the spores mature into free-living amebas.

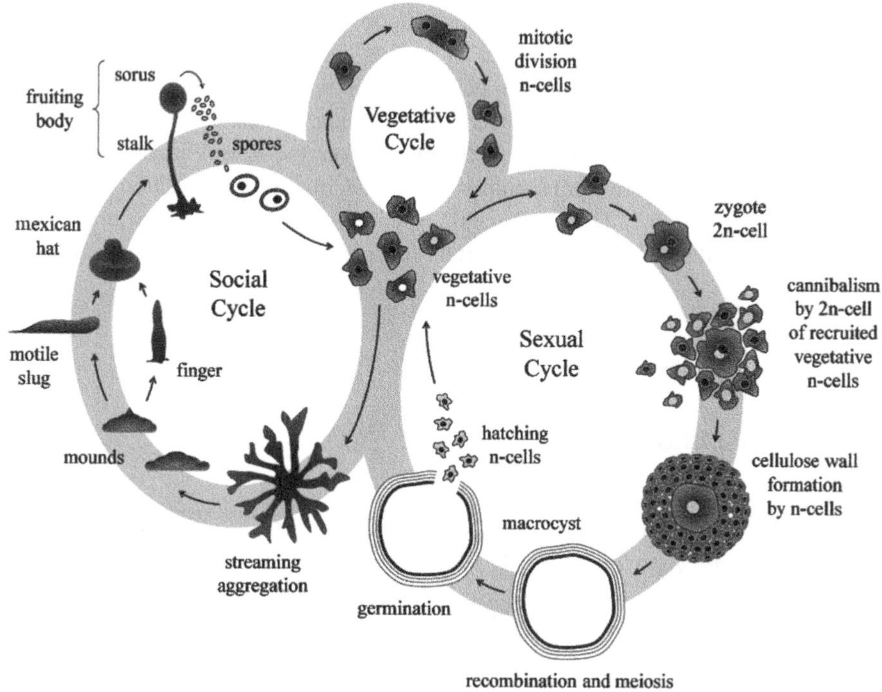

Figure 6.6. *Dictyostelium lifecycle (source: Paz-y-Miño-C and Espinosa 2018)*

Some authors recognize simply two-lifecycle stages in complex multicellularity: multicellular individual → zygote or equivalent (Grosberg and Strathmann 1998). Computer simulations (Biernaskie and West 2015) suggest that high relatedness facilitates the evolutionarily stable outcome (ESS = evolutionary stable strategy) of "high aggregation and high cooperation" because it makes it easier to surpass a threshold for cooperative aggregation to evolve. Such a threshold is taxa specific and environmentally influenced (António and Schulze-Makuch 2012). Since the benefits of cooperation must be shared among cooperators, the presence of unrelated cells, or cheaters, would discourage group formation due to genetic conflict. Thus, selection would favor above-threshold clonal numbers that overwhelm the presence of non-kin or cheaters (i.e. keeping them at low densities or eliminating them) and maximize the benefits of lasting multicellularity (Paz-y-Miño-C and Espinosa 2018).

Let us examine the micro anatomy of the fruiting body in *D. discoideum* (Figure 6.7). Remember that fruiting bodies are built upon recruitment of individual cells; these assemblages mature and differentiate according to function (Paz-y-Miño-C and Espinosa 2018). Division of labor in *D. discoideum* includes the formation of a protective "tip epithelium". Epithelia (epi = on top) are structured around specialized cells that cover the surfaces of bodies or organs. Epithelial cells are composed of "two sides" (= poles), one facing the outside of the body/organ, the apical side; the other, in direct contact with the subjacent cells of other tissues, the basal side (Paz-y-Miño-C and Espinosa 2018). The cell surfaces on the apical/basal sides of an epithelium form distinctive membranes as shown in the cross section of the tip epithelium in Figure 6.7. Thus, epithelial cells are "polarized" and their positioning is structurally and physiologically important for interacting with the environment.

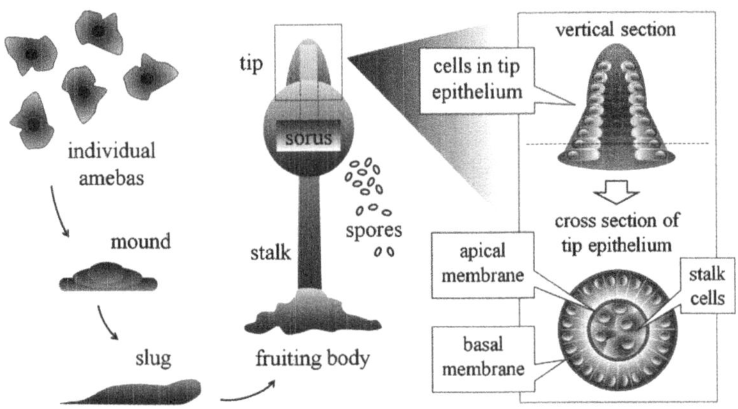

Figure 6.7. *Close-up of the social cycle in Dictyostelium, note the tip epithelium on the fruiting body (source: Paz-y-Miño-C and Espinosa 2018)*

As implied earlier, *D. discoideum* fruiting bodies are built upon recruitment of related amebas ($r \geq 0.9$). The cells continue development once joining the soon-to-be fruiting body (PCD affects ≈ 20% of cells; it takes place in the stalk and only amebas inside the sorus morph into spores; Queller *et al.* 2003). The reader shall recall that, according to inclusive fitness principles ($_rb - c > 0$), organisms maximize reproductive success by cooperating with kin and deterring unrelated cheaters. This premise has been tested in *D. discoideum* by running laboratory trials (involving mix, chimeric cell cultures of cooperators and mutant non-fruiting-body-producers = cheater "non-fruiters") and computer simulations in which the emergence of cheaters in a population was plotted as a function of the number of cell divisions that members of such population went through (Figure 6.8). This was based on the mutation rate that would take a cell to turn into a non-fruiter cheater, as well as the rate of cell proliferation that would allow a population to build a fruiting body (≈ 1000 to 100,000 cells, or ≈ 10–20 rounds of cell divisions; Kuzdzal-Fick *et al.* 2011).

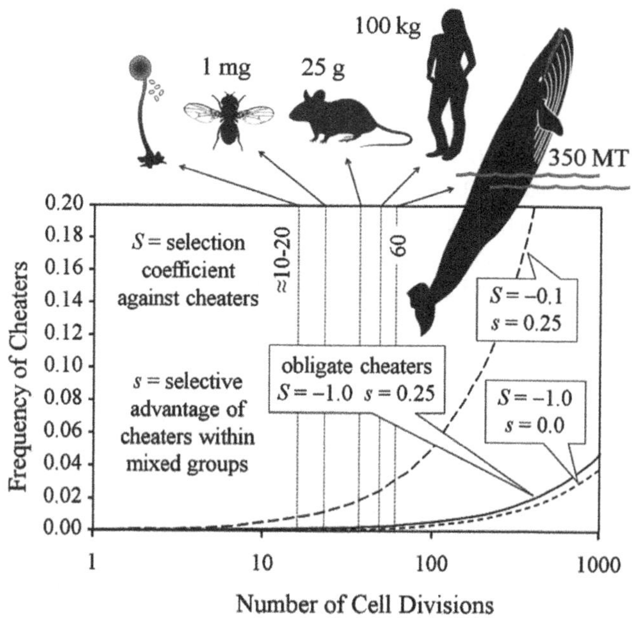

Figure 6.8. *Simulations of equilibrial frequency of cheater cells undergoing clonal development as a function of the number of cell divisions carried out by a founding cell (source: Kuzdzal-Fick et al. 2011; Paz-y-Miño-C and Espinosa 2018)*

The computer simulations included three scenarios (Figure 6.8): 1) a frequency growth of obligate non-fruiter cheaters in which the selection coefficient (*S*) against them was –1.0, at the same time that the selective advantage (*s*) on their side was 0.25 (solid line, Figure 6.8); this was conceivable for obligate cheaters, which ordinarily would not produce spores on their own and would depend on cheating ability to be 100% successful in turning into spores (best success they could accomplish was ≈ 80%). 2) a frequency growth of obligate non-fruiter cheaters in which *S* was also equal to –1.0, but *s* was 0.0 (short-dashed line, Figure 6.8); under these circumstances, the outcome was comparable to the previous simulation, although it did suggest that for simulations 1 and 2 mutation rate was the major source of cheaters. 3) a scenario in which *S* was very low at –0.1 (little selection against cheaters) and *s* was as favorable as 0.25 (long-dashed line, Figure 6.8); even under this advantageous scenario, obligate cheaters did not become a threat to single-celled-bottlenecked-based multicellularity (Kuzdzal-Fick *et al.* 2011).

The take-home message of Figure 6.8 is that the single-celled population bottleneck in the lifecycle of a multicellular eukaryote is so effective at deterring cheating that, at the rate of *D. discoideum* cell division, we could theoretically “build” not only the fruiting body of a social ameba (≈ 100,000 cells), but also a fruit fly (1 mg), a mouse (25 g), a human (100 kg), or a blue whale (350 MT); the latter would only require 60 rounds of cell divisions to be built (Kuzdzal-Fick *et al.* 2011)! Thus, the emergence of cheaters in a clonally-reproducing multicellular organism would never reach enough frequency to become a match to the fitness advantages of clonality and kinship (Kuzdzal-Fick *et al.* 2011; Paz-y-Miño-C and Espinosa 2018).

Some researchers reject the term cheater in chimeric assemblages and prefer to call these geno/phenotypes “loners”, with their own roles in the genetics of populations (Tarnita *et al.* 2015). Loners can be selected along with cooperators, persist in the community, and benefit from further selective advantage. Investigators have shown that experimentally induced low-genetic-relatedness in *D. discoideum* facilitates the coexistence of high rates (≈ 30%) of obligate-cheaters mixed with cooperators (Kuzdzal-Fick *et al.* 2011; Levin *et al.* 2015). These findings imply that a mechanism other than high relatedness (i.e. tolerance to cheaters) can sustain large cell aggregations over time (Paz-y-Miño-C and Espinosa 2018).

To close this section, we highlight that green-beard discrimination/recognition alleles have been characterized in *D. discoideum*. Briefly, to remain in proximity, amebas rely on cell membrane adhesion proteins encoded by *csA*. When wild-type *csA*$^+$ cells are grown in mixed soil cultures with *csA*$^-$ knockouts, the amebas cluster preferentially with those equipped with fully functional adhesion polypeptides (Queller *et al.* 2003; Paz-y-Miño-C and Espinosa 2018). Analogous experiments (wild-type vs knockouts) have been conducted with *tgrB1* and *tgrC1*, which also encode for cell adhesion transmembrane proteins (Benabentos *et al.* 2009; Paz-y-Miño-C and Espinosa 2016, 2018). However, distinctive from *csA*$^+$, which function is primarily adhesive, the *tgr* genes work in complementary pairs directly involved in cell-cell discrimination/recognition (Benabentos *et al.* 2009; Hirose *et al.* 2011; Strassmann and Queller 2011; Paz-y-Miño-C and Espinosa 2016, 2018). When fully functional pairs of *tgrB1*$^+$–*tgrC1*$^+$ are extracted from the wild (genetically different clones), expressed in identical cells which *tgr* genes have been previously knocked out (*tgrB1*$^-$–*tgrC1*$^-$), and grown in mixed cultures, the descendants segregate into distinctive mounds and fruiting bodies resembling the strains from which the wild-type genes originated (Hirose *et al.* 2011; Paz-y-Miño-C and Espinosa 2016, 2018).

6.5. Conclusion

Our current interest in understanding the many shapes of microbial detection of kin and kind can be, arguably, traced back to the 1850s and Darwin's curiosity in solving the "puzzle of the sterile social insects", in which the apparent sacrifice of the workers for the good of all could be explained by applying the mechanism of natural selection to the family. One hundred years later (1960s–1970s), Hamilton's concept of inclusive fitness and Maynard-Smith's postulate of kin selection offered the theoretical and mathematical basis to rationalize about how altruistic cooperation could turn into an ESS; that is, a balance between the benefits of helping another individual, the degree of genetic relatedness among cooperators, and the costs of helping ($_rb - c > 0$). Although the pioneering studies on discrimination/recognition of kin (1960s onwards) relied on work with multicellular eukaryotes (invertebrates, vertebrates, plants), work with microbes (2000s) produced genetic evidence (green-beard genes) that unicellular organisms could distinguish between same (clone) and different (diverse clones), as well as among conspecifics of gradual genetic proximity. Lately, studies with microbes have helped us understand the significance of the mechanisms of cell-cell discrimination/recognition for the advent of multicellularity (Paz-y-Miño-C and Espinosa 2016, 2018). Because microbes' lineages are ancient and occupy all environments, they are central to reexamining traditional hypotheses, reformulating concepts, and generating new knowledge on the evolution of kin and kind discrimination/recognition.

6.6. References

Adl, S.M. *et al.* (2019). Revisions to the classification, nomenclature, and diversity of eukaryotes. *Journal of Eukaryotic Microbiology*, 66(1), 4–119.

António, S., Schulze-Makuch, D. (2012). Toward a new understanding of multicellularity. *Hypotheses in the Life Sciences*, 2(1), 4–14.

Benabentos, R., Hirose, S., Sucgang, R., Curk, T., Katoh, M., Ostrowski, E.A., Strassmann, J.E., Queller, D.C., Zupan, B., Shaulsky, G., Kuspa, A. (2009). Polymorphic members of the lag gene family mediate kin discrimination in Dictyostelium. *Current Biology*, 19(7), 567–572.

Biernaskie, J.M., West, S.A. (2015). Cooperation, clumping and the evolution of multicellularity. *Proceedings of the Royal Society B: Biological Sciences*, 282.

Brunet, T., King, N. (2017). The origin of animal multicellularity and cell differentiation. *Developmental Cell*, 43(2), 124–140.

Cao, P., Dey, A., Vassallo, C.N., Wall, D. (2015). How myxobacteria cooperate. *Journal of Molecular Biology*, 427(23), 3709–3721.

Celiker, H., Gore, J. (2013). Cellular cooperation: Insights from microbes. *Trends in Cell Biology*, 23(1), 9–15.

Claessen, D., Rozen, D.E., Kuipers, O.P., Søgaard-Andersen, L., van Weze G.P. (2014). Bacterial solutions to multicellularity: A tale of biofilms, filaments and fruiting bodies. *Nature Reviews Microbiology*, 12(2), 115–124.

Darwin, C. (1859). *On the Origin of Species by Means of Natural Selection, or the Preservation of Favoured Races in the Struggle for Life*. John Murray, London.

Dawkins, R. (1976). *The Selfish Gene*. Oxford University Press, Oxford.

Dudley, S.A., Murphy, G.P., File, A.L. (2013). Kin recognition and competition in plants. *Functional Ecology*, 27(4), 898–906.

Espinosa, A., Paz-y-Miño-C, G. (2012). Discrimination, crypticity and incipient taxa in Entamoeba. *Journal of Eukaryotic Microbiology*, 59(2), 105–110.

Espinosa, A., Paz-y-Miño-C, G. (2014a). Evidence of taxa-, clone-, and kindiscrimination in protists: Ecological and evolutionary implications. *Journal of Evolutionary Ecology*, 28(6), 1019–1029.

Espinosa, A., Paz-y-Miño-C, G. (2014b). Examining crypticity in Entamoeba: A behavioral and biochemical tale. In *Why Does Evolution Matter? The Importance of Understanding Evolution*, Trueba, G. (eds). Cambridge Scholars Publishing, Newcastle, 181–190.

Espinosa, A., Paz-y-Miño-C, G. (2019). Discrimination experiments in Entamoeba and evidence from other protists suggest pathogenic amebas cooperate with kin to colonize hosts and deter rivals. *Journal of Eukaryotic Microbiology*, 66(2), 354–368.

Espinosa, A., Paz-y-Miño-C, G., Hackey, M., Rutherford, S. (2016). Entamoeba clone-recognition experiments: Morphometrics, aggregative behavior, and cell-signaling characterization. *Journal of Eukaryotic Microbiology*, 63(3), 384–393.

Fisher, R.A. (1930). *The Genetical Theory of Natural Selection: A Complete Variorum Edition*. Clarendon Press, Oxford.

Fisher, R.M., Cornwallis, C.K., West, S.A. (2013). Group formation, relatedness, and the evolution of multicellularity. *Current Biology*, 23(12), 1120–1125.

van Gestel, J., Tarnita, C.E. (2017). On the origin of biological construction, with a focus on multicellularity. *Proceedings of the National Academies of Sciences USA*, 114(42), 11018–11026.

Grosberg, R.K., Strathmann, R.R. (1998). One cell, two cell, red cell, blue cell: The persistence of a unicellular stage in multicellular life histories. *Trends in Ecology and Evolution*, 13(3), 112–116.

Haldane, J.B.S. (1932). *The Causes of Evolution*. Longmans Green, London.

Haldane, J.B.S. (1955). Population genetics. *New Biologist*, 18, 34–51.

Hamilton, W.D. (1963). The evolution of altruistic behaviour. *The American Naturalist*, 97(896), 354–356.

Hamilton, W.D. (1964). The genetical evolution of social behaviour. *Journal of Theoretical Biology*, 7(1), 1–16.

Herron, M.D., Borin, J.M., Boswell, J.C., Walker, J., Chen, I.-C.K., Knox, C.A., Boyd, M., Rosenzweig, F., Ratcliff, W.C. (2019). De novo origins of multicellularity in response to predation. *Scientific Reports*, 9(2328).

Hirose, S., Benabentos, R., Ho, H.-I., Kuspa, A., Shaulsky, G. (2011). Self-recognition in social amoebae is mediated by allelic pairs of Tiger genes. *Science*, 333(6041), 467–470.

Kaushik, S., Katoch, B., Nanjundiah, V. (2006). Social behaviour in genetically heterogeneous groups of Dictyostelium giganteum. *Behavioral Ecology and Sociobiology*, 59(4), 521–530.

Kraemer, S.A., Velicer, G.J. (2011). Endemic social diversity within natural kin groups of a cooperative bacterium. *Proceedings of the National Academies of Sciences USA*, 108, 10823–10830.

Kuzdzal-Fick, J.J., Fox, S.A., Strassmann, J.E., Queller, D.C. (2011). High relatedness is necessary and sufficient to maintain multicellularity in Dictyostelium. *Science*, 334(6062), 1548–1551.

Levin, S.R., Brock, D.A., Queller, D.C., Strassmann, J.E. (2015). Concurrent coevolution of intra-organismal cheaters and resisters. *Journal of Evolutionary Biology*, 28(4), 756–765.

Maynard-Smith, J. (1964). Group selection and kin selection. *Nature*, 201(4924), 1145–1147.

Maynard-Smith, J. (1977). Parental investment: A prospective analysis. *Animal Behaviour*, 25(1), 1–9.

Mehdiabadi, N.J., Jack, C.N., TalleyFarnham, T., Platt, T.G., Kalla, S.E., Shaulsky, G., Queller, D.C., Strassmann, J.E. (2006). Social evolution: Kin preference in a social microbe. *Nature*, 442(7105), 881–882.

Michod, R.E., Roze, D. (2001). Cooperation and conflict in the evolution of multicellularity. *Heredity*, 86, 1–7.

Mitri, S., Foster, K.R. (2013). The genotypic view of social interactions in microbial communities. *Annual Reviews of Genetics*, 47, 247–273.

Noh, S., Geist, K.S., Tian, X., Strassmann, J.E., Queller, D.C. (2018). Genetic signatures of microbial altruism and cheating in social amoebas in the wild. *Proceedings of the National Academies of Sciences USA*, 115(12), 3096–3101.

Ostrowski, E.A., Katoh, M., Shaulsky, G., Queller, D.C., Strassmann, J.E. (2008). Kin discrimination increases with genetic distance in a social amoeba. *PLoS Biology*, 6(11), 2376–2382.

Parks, D.H., Chuvochina, M., Waite, D.W., Rinke, C., Skarshewski, A., Chaumeil, P.-A., Hugenholtz, P. (2018). A standardized bacterial taxonomy based on genome phylogeny substantially revises the tree of life. *Nature Biotechnology*, 36(10), 996–1004.

Pathak, D.T., Wei, X., Dey, A., Wall, D. (2013). Molecular recognition by a polymorphic cell surface receptor governs cooperative behaviors in bacteria. *PLoS Genetics*, 9(11).

Paz-y-Miño-C, G., Espinosa, A. (2016). Kin discrimination in protists: From many cells to single cells and backwards. *Journal of Eukaryotic Microbiology*, 63(3), 367–377.

Paz-y-Miño-C, G., Espinosa, A. (2018). *Kin Recognition in Protists and Other* Microbes: *Genetics, Evolution, Behavior and Health*. Cambridge Scholars Publishing, Newcastle.

Penn, D.J., Frommen, J.G. (2010). Kin recognition: An overview of conceptual issues, mechanisms and evolutionary theory. In *Animal Behaviour: Evolution and Mechanisms*, Kappeler, P.M. (ed.). Springer, Berlin, 55–85.

Queller, D.C., Strassmann, J.E. (2013). Experimental evolution of multicellularity using microbial pseudo-organisms. *Biology Letters*, 9(1).

Queller, D.C., Ponte, E., Bozzaro, S., Strassmann, J.E. (2003). Single-gene green beard effects in the social amoeba Dictyostelium discoideum. *Science*, 299(5603), 105–106.

Ratcliff, W.C., Denison, R.F., Borrelloa, M., Travisano, M. (2012). Experimental evolution of multicellularity. *Proceedings of the National Academies of Sciences USA*, 109(5), 1595–1600.

Ruiz-Trillo, I., Nedelcu, A.M. (eds) (2015). *Evolutionary Transitions to Multicellular Life. Advances in Marine Genomics 2*. Springer, Berlin.

Schirrmeister, B.E., Antonelli, A., Bagheri, H.C. (2011). The origin of multicellularity in cyanobacteria. *BMC Evolutionary Biology*, 11(45).

Smith, J., Strassmann, J.E., Queller, D.C. (2016). Fine-scale spatial ecology drives kin selection relatedness among cooperating amoebae. *Evolution*, 70(4), 848–859.

Strassmann, J.E., Queller, D.C. (2011). How social evolution theory impacts our understanding of development in the social amoeba Dictyostelium. *Development, Growth & Differentiation*, 53(4), 597–607.

Tang-Martínez, Z. (2001). The mechanisms of kin discrimination and the evolution of kin recognition in vertebrates: A critical re-evaluation. *Behavioural Processes*, 53(1/2), 21–40.

Tarnita, C.E., Washburne, A., Martinez-Garcia, R., Sgro, A.E., Levin, S.A. (2015). Fitness tradeoffs between spores and non aggregating cells can explain the coexistence of diverse genotypes in cellular slime molds. *Proceedings of the National Academies of Sciences USA*, 112(9), 2776–2781.

Wall, D. (2016). Kin recognition in bacteria. *Annual Reviews of Microbiology*, 70(1), 143–160.

West, S.A., Griffin, A.S., Gardner, A., Diggle, S.P. (2006). Social evolution theory for microorganisms. *Nature Reviews Microbiology*, 4(8), 597–607.

Wilk, L., Strauss, M., Rudolf, M., Nicolaisen, K., Flores, E., Kühlbrandt, W., Schleiff, E. (2011). Outer membrane continuity and septosome formation between vegetative cells in the filaments of Anabaena sp. PCC 7120. *Cell Microbiology*, 13(11), 1744–1754.

7

Development and Evolution of Plant Forms

Florian JABBOUR[1] **and Guilhem MANSION**[2]

[1] *National Museum of Natural History, Paris, France*

[2] *Freelance biologist, Zurich, Switzerland*

7.1. Introduction

The delicacy of a wild impatiens corolla, the harmonious architecture of a mimosa leaf, or the all-geometric aesthetics of a grain of ipomoea pollen observed under the microscope… The myriad plant forms have always been – and still are – a source of astonishment, contemplation and even inspiration to which man cannot remain oblivious. As an integral part of the phenotype (from the Greek φαίνω [phaino], "to appear", and τύπος [tupos], "brand"), in other words, the set of micro or macroscopic characteristics observable in an individual, a form is determined by intrinsic (genetic background) and extrinsic (environmental conditions) factors. Its establishment during development (morphogenesis) is also modulated by physical and evolutionary constraints. Thus, the form of a plant tissue, organ or organism, as observed today, is the product of precise architectural and functional requirements.

In this chapter, we will first discuss the diversity of plant forms and the functions associated with them. We will then look at the development of these forms during the different stages of ontogenesis and phylogenesis of a plant. Finally, we will discuss the usefulness of plant forms, both for botanists whose aim is to classify nature and for human societies.

The Explosion of Life Forms,
coordinated by Georges CHAPOUTHIER and Marie-Christine MAUREL. © ISTE Ltd 2020.

7.2. Diversity of plant forms and associated functions

7.2.1. *Anthropocentric view of plant forms*

The forms of plants, as understood by scientists and users of plant resources, are parts of the phenotype discerned and described by man. This perception often differs from that of other animals because of the different structure of the eyes. Man's description of plant forms depends on his own powers of observation. It is based on qualifiers and measures that are significantly different from the perception of another mammal, insect, bird, microscopic organism or animal whose vision is only effective in ultraviolet light.

The scientific endeavor to describe plant forms is generally based on their prior categorization and classification. By applying a theoretical grid of group ordering to the diversity of plant forms, following concepts such as symmetry (Citerne *et al.* 2010; Damerval *et al.* 2017) or modularity (a type of organization where units or "modules", sometimes redundant, are more connected to each other than to other structures of the biological system) (Diggle 2014), forms become objects that can be more easily analyzed and compared. Form, as a concept perceived and categorized by humans, can thus characterize various, sometimes nested, levels of organization: the silhouette of the crown of a tree, the organization of an inflorescence, the branching pattern of a root system, the contour of a leaf, the modular profile of a legume fruit, the symmetry of a flower, etc. (Diggle 2014).

In addition to the direct observation of macroscopic forms, which is almost a thousand years old, man now has access to the infinitely small. The progress made over four centuries in the manufacture of optical instruments (such as binocular magnifiers and scanning microscopes) now makes it possible to routinely describe the scaled surface of the tissue of a petal or to detail the chronological sequence of organ development in a flower bud, among other examples.

The plant forms that are thus accessible to man are fascinatingly diverse, as shown by the multiplicity of vegetative and reproductive organs observed in plants (Rousseau 1780–1789; Darwin 1877; Endress 1996), some examples of which are developed in this chapter. In addition to this immense interspecific variation, significant morphological differences can sometimes be observed within the same species, or within the same individual, and result from the influence of external factors such as climate (this is known as "phenotypic plasticity", which refers to the capacity of an organism to present a modified morphology and physiology, in relation to changes in environmental conditions) or intrinsic variables (age, position of the organ, etc.) (Gould 1993). For example, in love-in-a-mist (*Nigella damascena* [Ranunculaceae]), plants with a perianth that is composed of similar organs (tepals) coexist and reproduce with plants whose flowers have a perianth differentiated into sepals and petals (Damerval *et al.* 2018). In some species of Geraniaceae or Malvaceae, flowers with dextrorse and sinistrorse organ arrangements are sometimes observed within the same individual (Endress 1999).

Lastly, we can add aberrant or monstrous forms to our catalogue of normal (or typical) forms, which may result from irregularities in the expression of developmental genes. Examples include the teratological floral forms of orchids (Mondragón-Palomino and Theißen 2009) or the modified forms of cultivated roses in which the floral organs are replaced by leaves (Yan *et al.* 2016).

7.2.2. *Plant forms perceived by pollinators*

The various forms (contours, symmetry, etc.) of a plant structure, as perceived by animals, is probably not the same one that could be described by a human observer (Krishna and Keasar 2018). Thus, pollinators are often sensitive to signals that can only be observed in ultraviolet light, such as "nectar guides" (lines detaching from the dark background and leading to a sweet reward that is produced and/or stored in a pocket or a spur), or to certain structures invisible to the human eye, such as nanoscales that are present on various floral organs, which have anatomical and optical properties that generate a visual signal that can be perceived by insects (Moyroud *et al.* 2017).

Regarding floral morphology, an insect will generally only perceive the general symmetry (Møller 1995), whereas the botanist will be able to describe it finely by a thorough quantification of the different degrees of asymmetry (Gómez *et al.* 2006). Thus, deviations from perfect symmetry, whether bilateral or radial, are sometimes observed in floral development studies (Jabbour *et al.* 2009). Indeed, when the sepals and petals are initiated according to a spiral phyllotaxis (arrangement of the organs on one axis) on the floral meristem (undifferentiated tissue), the symmetry of the flower, as understood in the mathematical sense of the term, in a two-dimensional plane, cannot be considered perfect.

The forms perceived by the pollinator (generally approaching an anthetic flower head-on) and the botanist (studying flower buds and flowers using microscopy, anatomy and morphometric techniques) are therefore sometimes quite different. In ecology studies that examine the relationship between plants and their biotic environment, it is therefore important not to consider human perception of forms as a working hypothesis, but to be aware that the forms to be analyzed are probably very different from those we can describe.

7.3. Origin and evolution of plant forms

7.3.1. *Pattern formation during ontogenesis*

Goethe (1790) was one of the first to conceptualize the study of plant morphology and to not only give it a descriptive, but explanatory aspect too. According to him, the different parts of a flower result from the metamorphosis (meant here in the sense of transformation) of a series of organs that he called "ideal leaves".

In general, the form of an organ is described once it has reached maturity. While the development of an organ is often documented to understand its morphogenesis (see Figure 7.1A, D), the stages corresponding to the evolution of the morphology of the so-called adult organ, up to its senescence (e.g. Figure 7.1E, F), are rarely documented. These last stages are, however, very easy to observe and provide a great deal of information on the formation and constitution of the organ.

In flowering plants, the meristem (from the Greek μέρος [*meros*], "portion"), a tissue composed of undifferentiated stem cells, is responsible for the formation and growth of new tissues (histogenesis) and organs (organogenesis). The meristem thus allows the creation of a wide variety of organic assemblages, a true product of the initial position of the islets of organogenous cells, their order of initiation or their number. Organogenesis and morphogenesis events (developmental events that participate in the acquisition, by a structure, of its adult form) correspond to deformations of tissues localized at specific locations of the meristem and organs, respectively. Cell division and differentiation phenomena are determined by the action of proteins and hormonal signals within the meristem and the primordia, which are the organs in their earliest developmental stages (Citerne *et al.* 2010, Damerval *et al.* 2017). Auxin and the AHP6 protein have been shown to have antagonistic actions, respectively activating and inhibiting plant organogenesis (Besnard *et al.* 2014). Auxin plays a major role in establishing the type of phyllotaxis in *Arabidopsis thaliana*, the organs being initiated at locations corresponding to the local maxima of production of this hormone.

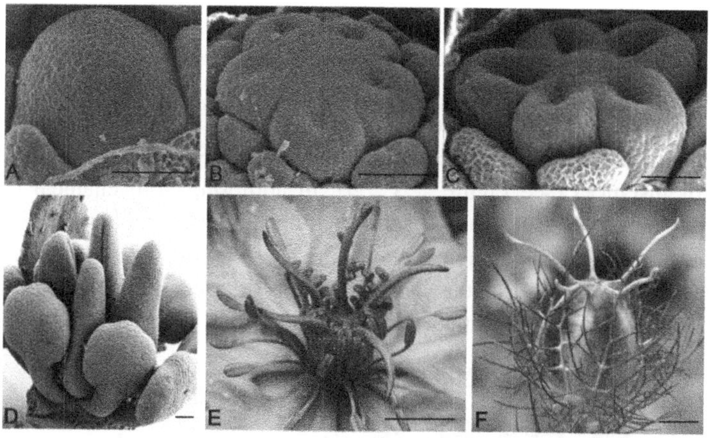

Figure 7.1. *Form of the gynoecium of the love-in-a-mist flower (Nigella damascena, Ranunculaceae). For a color version of this figure, see www.iste.co.uk/chapouthier/life.zip*

COMMENT ON FIGURE 7.1.– *A: gynoecium not yet present on the surface of the flower meristem; B: initiation of the five carpellar lodges; C and D: development and growth of carpels; E: carpels in a flower at anthesis; F: fruit (capsule). A-D: photographs taken with a scanning electron microscope. Scales: A-D: 100 μm; E-F: 1 cm (© Florian Jabbour).*

The study of genes influencing the evolution and development of structures and forms, also known as "evo-devo", has led to enormous progress in understanding the processes of organogenesis and morphogenesis (Coen *et al.* 2017; Moyroud and Glover 2017; Rebocho *et al.* 2017a, 2017b). Thus, Goethe's initial hypothesis of "metamorphosis" has been validated by the ABCE model, which states that a set of molecular interactions between different classes of genes (A, B, C, and E) allows the identity of flower organs to be determined (Theissen and Saedler 2001; Irish 2017). The cadastral genes, expressed upstream of the A–E genes, participate in the delimitation of the territories of the floral meristem that will be occupied by different types of organs. Other genes (*CUC*, *CYC*), located downstream of genes A–C and E, ultimately allow the expression of the aspect and form of the newly formed organs.

7.3.2. *Physical-mathematical considerations on plant morphogenesis*

The caulinary apical meristem sequentially generates modules, or phytomers, composed of a leaf, an axillary bud and an internode. Similarly, an inflorescence meristem generates flowers and a floral meristem generates floral organs. The phyllotaxis of the floral organs may be whorled or spiral. In the latter case, the angle of divergence between two consecutive organs approaches 137.5° (the golden angle) (see Figure 7.2A, B). The organs are as if inserted on imaginary curved lines (parastichies) and the number of parastichies, counted clockwise and counterclockwise, are two consecutive numbers of the Fibonacci sequence (8–13 or 13–21, for instance) (see Figure 7.2C, D). In addition, biomechanical constraints (Sassi and Traas 2015) influence organ initiation. Experiments have shown that the initiation of organs on a meristem follows a relatively simple physical principle: the youngest organ is formed at the location where the occupation of the previously initiated organs was the least important (Douady 2011).

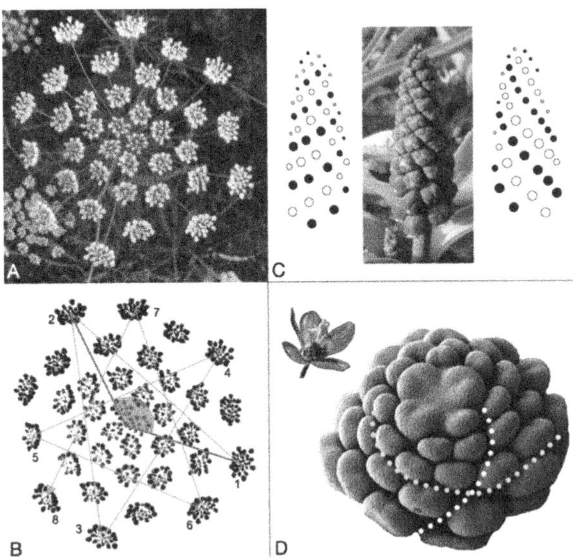

Figure 7.2. *Phyllotaxis of the reproductive organs of some flowering plants at different levels of organization: inflorescence of inflorescences, inflorescence and flower bud. For a color version of this figure, see www.iste.co.uk/chapouthier/life.zip*

COMMENT ON FIGURE 7.2.– *A: an umbel of umbels of an Apiaceae; B: black-and-white version of the same inflorescence, showing the first eight umbels initiated on the inflorescence meristem. The angle between two successive umbels (in blue) is about 137.5°; C: inflorescence of Muscari armeniacum (Asparagaceae). Two series of parastichies, clockwise (drawing on the right) or counterclockwise (on the left), can account for the insertion of the flowers on the axis of the inflorescence; D: flower bud of Staphisagria picta (Ranunculaceae, adult flower on the left) at the stage of carpel initiation (here, three) at the top of the meristem. The primordia below are stamen primordia. The flat organs at the base of the meristem are perianth organs. Two parastichies are shown in yellow (clockwise) and white (counterclockwise) (© Florian Jabbour).*

7.3.3. *Implementation of forms during phylogenesis*

In order to trace the course of the evolution of forms, two types of information are needed: 1) a description of the different states for the morphological traits studied in related lineages and 2) a phylogenetic tree representing the evolutionary proximity between the current (sometimes fossil) lineages. The reconstruction of the evolution of ancestral states at each node of the tree (going back to the hypothetical common ancestor) is done using probabilistic methods that take the degree of kinship of the lineages into account. These inferred ancestral forms can then be compared to fossil forms.

The different evolutionary processes (mutation, genetic drift, gene flow, natural selection) explain the great diversity of observed forms, their maintenance or disappearance during evolution, as well as some convergences (similar forms observed in unrelated evolutionary lineages) (Chomicki *et al.* 2017). A global study, on the scale of all flowering plants, has thus made it possible to reconstruct the probable portrait of the ancestral flower, a hypothetical combination of states of traits for each type of floral organ (number and forms of carpels, stamens and perianth parts). However, it is surprising to note that the combination of characteristic states observed on this protoflower is not found in the present species nor in known fossil species (Sauquet *et al.* 2017; Sokoloff *et al.* 2018).

A powerful way of highlighting and comparing the forms achieved during evolution is the construction of morphospaces (Chartier *et al.* 2014, 2017). A morphospace is a theoretical mathematical space with n-dimensions (often restricted to two or three dimensions for ease of reading) that describes and connects the different phenotypes in a lineage. The construction of such a space is made possible by the ordination of a set of particular variables (in this case, morphological traits). These morphological spaces make it possible to distinguish phenotypes realized during the evolution of a lineage from phenotypes that are theoretically possible but never observed, due to developmental, functional or phylogenetic constraints.

7.3.3.1. *Forms of vegetative structures and associated functions*

The forms of the organs of the plant's vegetative apparatus may indicate adaptation to the surrounding environment and the constraints it imposes. Anchoring to the ground and the capture of mineral resources will be carried out differently depending on whether the root is from the taproot type or fibrous. Underground and aerial organs in certain groups of plants show surprising adaptations to aridity. Water is stored in these structures, which are therefore swollen, as is the case with Malagasy euphorbias (Evans *et al.* 2014). Regions with arid climates are also home to plants that are sometimes much smaller than those of related species living in other less constraining climates. In *Delphinium*, from the Iranian-Turanian region, for example, individuals are frail, short, and the leaves are thin and few in number (Jabbour and Renner 2011). The correlation between leaf form and climate has been demonstrated in several cases (Peppe *et al.* 2011). Entire, waxy and broad leaves are often associated with tropical climates, leaves with leaflets or reduced to spines are typical of hot and arid environments.

These differences in form and texture are related to different performances of photosynthesis and/or control of water losses by evapotranspiration. In some rarer cases, the vegetative parts of the plant are adapted to a less common mode of nutrition: that of a carnivore. Structures in the form of an urn (*Nepenthes*) or a flytrap (*Dionaea*) that attract insects in search of food or shelter capture them and "digest" the fluids produced by plant enzyme actions on the trapped bodies – these are the structures that have been selected over the course of evolution (An *et al.* 2002; Libiaková *et al.* 2014).

In addition to form, the arrangement of leaves (phyllotaxis) is also subject to extreme variation, with little understanding of its "usefulness". Thus, the hypotheses advocating optimal light uptake or protection of young organs from environmental stress have never been clearly demonstrated. Another theory, which is not adaptationist, would suggest that phyllotaxis is simply the result of the way in which organs are initiated and develop, two phenomena that depend on molecular interactions within cells and physical constraints within the meristem (Adler *et al.* 1997; Newell *et al.* 2008).

7.3.3.2. *Forms of reproductive structures and associated functions*

The most striking floral element for the human eye is perianth. It can be differentiated into a set of sepals (calyx) and petals (corolla), or only made up of undifferentiated tepals. Diversity in perianth organization comes from the combination of multiple states of traits such as : 1) radial symmetry or actinomorphy (see Figure 7.3A, E, F, H, I); 2) bilateral symmetry or zygomorphy (see Figure 7.3B, C, D, G, J, K); 3) general campanulate form (see Figure 7.3A), tubular (see Figure 7.3B, C) or bilabiate (see Figure 7.3D); 4) the total or partial fusion (see Figure 7.3A, B, J) of the petals; 5) the presence of appendages (see Figure 7.3K, nectariferous spur of a tepal). These forms are produced during the development of flower buds and are the result of evolutionary processes in which the selection of morphologies that ensure optimal pollination plays a major role (Endress 1999; Zhong and Preston 2015).

Figure 7.3. *Diversity of floral forms in angiosperms. Some examples of perianths classified according to the Tournefort model, where classes 1 to 15 (A-O) include all herbaceous and suffrutescent plants. For a color version of this figure, see www.iste.co.uk/chapouthier/life.zip*

COMMENT ON FIGURE 7.3.– *A: Campanula barbata (Campanulaceae, class 1); B: Lycium intricatum (Solanaceae, class 2); C: Aristolochia baetica (Aristolochiaceae, class 3); D : Antirrhinum charidemi (Plantaginaceae; class 4); E: Heliophila coronopifolia (Brassicaceae; class 5); F: Cistus albidus (Cistaceae; class 6); G: Orlaya grandiflora (Apiaceae; class 7); H: Silene colorata (Caryophyllaceae; class 8); I: Moraea tricolor (Iridaceae; class 9); J: Lathyrus clymenum (Fabaceae; class 10); K: Anacamptis papilionacea (Orchidaceae; class 11); L: Adenostyles leucophylla (Asteraceae; class 12); M: Pilosella aurantiaca (Asteraceae; class 13); N: Felicia sp. (Asteraceae; class 14); O: Lygeum spartum (Poaceae; class 15) (© Guilhem Mansion).*

It is easy to conceive that the various forms of fruit promote different ways of dispersing and disseminating seeds. *Papaver* seeds escape through the openings at the top of the wind-swirled capsule (see Figure 7.4A). Other capsules open, when dry, and the seeds simply fall to the base of the mother plant (see Figure 7.4B). The maturation of the ovary into fruit is sometimes accompanied by the transformation of the associated floral tissues, such as those of the flower receptacle (see Figure 7.4C, D, F), which may then perform functions to protect the fruit or, if fleshy and consumed by animals, to facilitate seed dispersal. Fruits of Fabaceae (legumes) have two valves that abruptly detach from each other at maturity, resulting in passive propulsion of the seeds on the ground, at a short distance from the mother plant (Figure 7.4E). In contrast, the seeds of the squirting cucumber (*Ecballium elaterium*, Cucurbitaceae) are propelled relatively far by a process of explosive dehiscence of the fruit. Other fruits have a form that allows them to float until they meet a stretch of shoreline where the seed can germinate. The fruits of *Barringtonia* (Lecythidaceae) in the form of a bishop's cap are a perfect illustration of this (see Figure 7.4H). The samaras are carried by the wind some distance from the mother plant by the winged extension of the fruit tissues (see Figure 7.4I). Fruits are sometimes grouped into infrutescences (see Figure 7.4G), and the form of the infrutescences can provide criteria for distinguishing species within a group.

Beyond the macroscopic structures, the extreme diversity of forms and microscopic structures observed in pollen grains has allowed the establishment of a particular science: palynology. This diversity of phenotypes can be related to the modes of transport of pollen grains, as well as to the different types of abiotic environments in which pollen grains can germinate. Thus, the more or less complex ornamentation of the exine (wall of the pollen grains made up of a biopolymer close to chitin, sporopollenin) could be linked to the diversity of pollinators on which the pollen clings (Banks and Rudall 2016). The form of some pollen grains with air sacs can facilitate their transport by wind. Lastly, particular mesh sizes of the exine on the surface of the pollen grain may be able to withstand high levels of dehydration (Matamoro-Vidal *et al.* 2016) before encountering a stigma and germinating there.

Figure 7.4. *Diversity of fruits illustrated by some samples kept in the herbarium of the National Museum of Natural History in Paris. The name of the collector and the collection number are specified when the information is available. For a color version of this figure, see www.iste.co.uk/chapouthier/life.zip*

COMMENT ON FIGURE 7.4.– *A: capsules of Papaver somniferum (Papaveraceae), Paillot 2407, collected in 1858 in France; B: capsule and seeds of Helicteres (Malvaceae), Dupré 87, collected in Brazil; C: achenes of Nelumbo nucifera (Nelumbonaceae) in the transformed fleshy floral receptacle, Douillot s.n., collected in 1883; D: capsules of Eucalyptus kingsmillii (Myrtaceae) in the processed floral receptacle, Craven 5158, collected in 1978 in Australia; E: pod of Albizia lebbek (Fabaceae), collected in 1878 in Egypt; F: polyfollicle of Banksia grandis (Proteaceae), leg. ign. 476, collected in 1845 in Australia; G: capsules of Symingtonia populnea (Hamamelidaceae) organized in an infrutescence, leg. ign. 1269, collected in 1995 in Vietnam; H: capsule of Barringtonia speciosa (Lecythidaceae), Decary 14650, collected in 1939 in Madagascar; I: samara of Centrolobium robustum (Fabaceae), Glaziou 6221, collected in 1887 in Brazil. Scale: 1 cm (© Florian Jabbour).*

7.4. Origin and evolution of plant forms

7.4.1. *Usefulness for human societies*

7.4.1.1. *The recognition and oral communication of plant forms*

From an early age, human societies have been confronted with a multitude of surrounding forms, whether it be animal, mineral or plant. The unconscious recognition of a certain number of them must have very quickly led to the existence of empirical knowledge, transmitted orally from generation to generation. The first step was to differentiate between "useful", edible, curative or decorative plants and harmful, malignant and poisonous plants. The memorization of the external appearance, colors, smells and flavors thus meant the recognition of the plant object: recognizing the form became a vital act! This selective and instinctive sorting augured, as it were, the beginnings of an empirical plant classification.

7.4.1.2. *Towards a written transmission of plant forms*

The plant form, once recognized, could have been celebrated, even idealized or divinized, for its many virtues, as evidenced by the plethora of vegetal subjects found in Egyptian wall paintings, Assyrian bas-reliefs, Minoan frescoes and other Greek vases. These engraved, highly stylized representations can today be seen as references for a better identification of plants. This need for visual recognition of forms led to the publication of the first botanical illustrations in the *Rhizotomikon* of Crateus (around 120–60 BCE), drawings that were abundantly copied and that would endure in the rare illustrated copies of Dioscorides' *De Materia Medica* ("Dioscorides of Vienna", around 515 CE). Despite their apparent simplicity, these plant representations were invaluable to medical practitioners in the Middle Ages. A drawing is worth more than a thousand distorted or misinterpreted words!

7.4.1.3. *Towards an interpretation of plant forms: doctrine of signatures*

The dark ages following the ephemeral advent of philosophical and critical Greek botany (about 4th Century BCE), a more materialistic and slavery-centered Roman period, and then a stranglehold of monotheistic religions on the sciences, led to a general stagnation of thought, with the development of many esoteric doctrines and other fanciful beliefs, to say the least. It is in this ambient obscurantism that the famous "doctrine of signatures" was developed, among other things, which states that every plant (and by extension every living, even inert, being) expresses, by its external aspect, its hidden properties, its invisible reality, which every good naturalist must interpret. Plant organs in the form of a liver (lobed leaves of *Hepatica nobilis*), a lung (spotted leaves of *Pulmonaria officinalis*), a heart (pepper, tomato), a kidney (beans) or a brain (walnut kernels) are all remedies for liver, lung, heart, kidney or brain diseases. The underground parts are not to be ignored: the roughness of the twig root is reminiscent of hemorrhoidal inflammation; the tubers of *Orchis* justify their Greek name "testicles"; the human form attributed to the roots of mandrake or ginseng (*Panax*) makes them universal remedies or "panaceas", the latter term deriving directly from *Panax*. The color of vegetable juices should make you think of body fluids: bile yellow, blood red, milky or spermatic white, etc. We could multiply the examples at our leisure! This very anthropocentric way of apprehending plant forms, proposed by Aristotle under the name of "physiognomy" and taken up by certain botanists such as Brunfels (1530) or Fuchs (1543), reached its climax with the Swiss physician Paracelsus (1571) and the Italian writer Della Porta (1588). The latter spent his life searching for a sign of hidden power or personality in every living form, including man. At best, "signatures" could be considered as a mnemonic, rather naïve, means of recognition of plants. And, to quote the biologist Guyenot (1941): "All this is extremely childish and marks a real step back from the extremely serious work of the botanists of the same period."

7.4.1.4. *Towards a copy of plant forms: bionics or biomimetics*

If man has not always been very happy with his interpretation of plant signatures, he has nevertheless been able to find a more material inspiration in the multitude of surrounding forms. This copying of biological models, whether structures or mechanisms, for technological purposes is the very essence of "bionic" or "biomimetic" science, the name officially given to it at a scientific congress in Dayton, Ohio, in 1960. This date marks the beginning of multidisciplinary research, bringing together biologists, chemists, engineers, mathematicians and physicists, with the aim of designing efficient artificial systems derived from the observation of natural forms. In this field, plants have proved to be a prolific source of ideas, to say the least.

For example, one of the first bionic patents is still found on the American continent. The barbed wire, or "devil's horns", developed by Joseph Farewell Glidden (1874), derives directly from the copy of the thorny branches of a local shrub, *Maclura pomifera* (Moraceae). The latter was used by the local Sioux tribes, the Osages, to make weapons, war paintings (hence the vernacular name "Osage orange"), but above all, impassable hedges for animal enclosures.

On the other side of the Atlantic, architect Joseph Paxton was inspired by the leaf structure of a tropical water lily, the *Victoria amazonica* (Nymphaeaceae), to make the roof of London's famous Crystal Palace for the 1851 Great Exhibition. Noting that the combination of rigid radial ribs and flexible concentric ribs formed a light and extremely strong leaf framework capable of supporting the weight of his eight-year-old granddaughter, Paxton drew the plans for a building in a few days, one that would stand the test of time and whose only weakness would be fire.

Another famous case of ingenious invention derived from the observation of plant forms is to the credit of the Swiss engineer Georges de Mestral. He wanted to remove the numerous burdock fruits (*Arctium lappa*, Asteraceae) that had been left hanging from his dog's fur during a hunting trip, and was intrigued by their structure. He decided to observe them carefully under a microscope and realized the presence of small special deformable hooks that allow extremely strong, but reversible, connections with smooth fibers. The resulting industrial copy (from Mestral 1961), combining a plain, velvet surface and a rough strip of hooks, would bear the hybrid French name "velours-crochet" or "Velcro®".

Similarly, the observation of poppy capsules and the way they spread the seeds inspired the engineer Raoul Francé, who made salt shakers and other similar objects allowing a very homogeneous dispersion of powders (Francé 1919). There are many examples of copies of patented plant forms: pappus seeds of *Taraxacum*, *Tragopogon* (Asteraceae) or *Valeriana* (Valerianaceae), reminiscent of small parachutes; hydrophobic surfaces of certain leaves of agaves, coffee trees or other fountain trees (*Ocotea foetens*, Lauraceae); without forgetting simpler structures such as the fluted stems of umbelliferous plants, which undoubtedly inspired the columns of ancient temples.

7.4.2. *Usefulness for botanical classifiers*

7.4.2.1. *From the amorphous to a specific identity of the plant form*

Although the usefulness and beauty of plants have been known since the dawn of time, it was not until classical antiquity that the identity of the plant kingdom and its representatives, considered up until then to be purely decorative elements, began to be understood. The first Greek thinkers, led by Aristotle, considered all living organisms as derivatives of primitive initial forms, mixtures of water and earth, gradually modified to better adapt to their environment. Plants would only be bisexual and fixed animals, feeding by their roots and propagating by their seeds.

7.4.2.2. *First studies of plant forms: the advent of "phytology" or botany*

It was not until the Greek philosopher Theophrastus' *Historia Plantarum*, between the 4th and 3rd Centuries BCE, that a clear separation between the animal and plant kingdoms appeared, as well as a first attempt at a written classification of plants (Theophrastus 1988). This disciple of Aristotle approaches the study of plants in a scientific manner for the first time and proposes to group the 500 or so plants that were then known, whether wild or cultivated, into four distinct classes: trees, shrubs, sub-shrubs and herbs. From there, he proposes a number of useful characteristics that would enable him to refine his diagnosis. In particular, he highlights the distinctions between mono- and dicotyledons, sepals and petals, corollas with free and fused petals, actinomorphy and zygomorphy. He was also the first to be interested in the general morphology of plants, their growth according to external conditions and their mode of reproduction. In particular, he postulates that each plant bears both male and female sexes, even if he does not explicitly distinguish between them, and recognizes the transformation of carpels into fruits containing seeds. Although he prefigures all the theoretical aspects of modern botany, from anatomy-morphology to systematics, including physiology, ecology and biogeography, he does not forget the applied side of modern botany by rigorously describing the economic and pharmaceutical properties of plants and their derivatives (gums, tars, resins or juices).

7.4.2.3. *Usefulness before form: botany distorted and phagocytized by medicine*

This fundamental work, constantly revised by a philosopher who was always questioning himself, would unfortunately be nothing but a flash in the pan. The followers of Theophrastus, compilers for the most part, indeed only saw a source of potential medicines in plants, inevitably reducing the young botanical science to a branch of medicine. The work of the "father of botany" – and thus the study of the classification of plant forms – would therefore have to wait more than 18 centuries and past the early Renaissance before being revived. Thus, only the writings of Dioscorides (*De Materia Medica*) or Pliny the Elder (*Naturalis Historia*) made their way through the Middle Ages, either through patient Latin copies from Christian monasteries, or through Arabic translations in Muslim lands, where their late discovery seemed to give them a new lease of life. In any case, both Western and Eastern versions remain for the most part confusing, distorted and, above all, devoid of any critical sense.

7.4.2.4. *The rebirth of plant forms: a revival of realistic and critical botany*

The beginning of the Renaissance marked a kind of break with the medieval sclerotic spirit with, in particular, a rediscovery of Greek philosophical thought via the first Latin editions of Theophrastus or critical translations by Dioscorides. Little by little, plants would be studied as independent organisms; in other words, one would be as interested in their forms as much as their curative properties. The appearance of the first botanical gardens, the development of printing, and the advent of a certain artistic realism were all factors that favored the production and dissemination of new, purely botanical and richly illustrated works. The main figures of these revolutionary works, in which the reality of forms and subjects took precedence over their divine or medical symbolism, were Brunfels (1530), Fuchs (1543), and Bock (1546). The appearance of paper, in addition to its primordial role in the production of books, also contributed to the creation of the first herbaria. This invention, attributable to the Italian botanist Ghini, allowed, among other things, an almost perfect conservation of plants in the form of bidimensional objects that could be observed at leisure (Isely 2002).

7.4.2.5. *The growing diversity of plant forms: towards a necessity for classification*

The great naval explorations and the appearance of the first botanical gardens saw an exponential increase in the number of known plant species over the following centuries, leading to the imperative need for accurate botanical descriptions and effective classification systems. It was absolutely necessary to catalogue this plant chaos and group the forms into recognizable entities!

Cesalpino was the first to propose a grouping based on the structure of the flowers and fruits, as a certain Theophrastus once did. His *De plantis libri* lists some 1,500 plants and heralds the notion of genus (Cesalpino 1583). Bauhin goes a little further and tries to divide 6,000 plants into 72 sections, according to a set of common morphological characteristics. His major works prefigure the binomial nomenclature, as well as the concepts of genera and species (Bauhin 1623). Ray refined the work of his predecessors in his *Historia Plantarum* (Ray 1686), aided by a precise botanical terminology published by Jung (1747). In particular, he is credited with a better demarcation of taxonomic entities above the genus level and the first formal distinction between monocotyledons and dicotyledons (Morton 1981). The work of Ray particularly influenced Magnol, who officially published the first concept of family, as well as his pupil, the insatiable traveler Tournefort. The latter would see the basic unit of botanical classification in the genus (*essentiae*), from which the species (*differentiae*) naturally derives. Tournefort also presents a so-called "artificial" classification in his *Institutiones Rei Herbariae* (1700), in other words, a classification based on a few subjectively chosen characteristics. Based mainly on floral criteria (corollary form, petal fusions, ovary position), but ignoring the cryptogamous/phanerogamous or monocots/dicots divisions of its predecessors, he groups more than 7,000 species into some 700 genera and 22 classes (see Figure 7.4). His system, simple to use and richly illustrated by Aubriet, was a great success and influenced future generations, including a certain Linnaeus. The latter, a tireless compiler, standardized the concepts of genera and species dear to Ray and Tournefort (Linnaeus 1735, 1737), and generalized the binomial nomenclature, initially proposed by Bauhin two centuries earlier (Linnaeus 1753). Very much influenced by Vaillant's provocative theory on the sexuality of plants (1718), he developed a new artificial classification (called "sexual"), based on the number and position of male (stamens) and female (pistils) genitalia. This new system, although impractical for the identification of species, was widely adopted and concluded the Swedish botanist's success.

7.4.2.6. *Towards the implementation of natural morphological classifications*

Paradoxically, Linné's work marked the end of artificial systems, which were too subjective. The time had now come for natural classifications. The latter took all possible forms into account – without a priori – and ultimately made it possible to identify the states of traits that best described the different taxonomic groups. Initially proposed by Ray and introduced in France by Magnol, this concept of natural systematics was widely taken up by the Jussieu dynasty before reaching its peak with Adanson's *Familles des plantes* (1763). It is the apogee of morphology: all possible forms are used to classify plants according to their total resemblance.

7.4.2.7. *Death to classical morphology... long live evolutionary morphology!*

With the birth of evolutionary thought, initiated by Lamarck (1809) and developed by Wallace (1858) and Darwin (1859), this search for total similarity that we would describe today as "phenetic" will gradually give way to classifications named "cladistic" or "phylogenetic". Plant species change over time and it is now important to reconstruct their phylogeny by inferring their relationships of evolutionary proximity. Moreover, the immense technical progress made in biological investigation (molecular biology, cytology, phytochemistry) provides access to taxonomic traits that are much more stable than the different forms, whose variation is not always genetic. A large number of phylogenetic systems have been published, from Engler's *Syllabus der Planzenfamilien* (1898) to the first molecular version of the *Angiosperm Phylogeny Group* (1998), via Cronquist's classification (1981).

At the beginning of the 21st Century, it must be noted that, in the vast majority of cases, morphological characteristics are no longer used a priori in the elaboration of plant classifications. On the other hand, the study of their evolution in light of a phylogenetic pattern has opened the door to a very large number of new perspectives: let us mention, among others, the distinction between synapomorphic and homoplasious traits in systematic botany; the understanding of morphological complexity in evo-devo; the development of phyllotactic theories, etc. The tool used since the dawn of time has now become a fascinating object of study.

7.5. Conclusion

The immense morphological diversity of plants continues to attract the ever-growing interest of biologists. It also arouses the curiosity of a wider scientific community fascinated by various aspects of plant forms. Indeed, modern studies of plant morphology can no longer be carried out without recourse to disciplines as varied as evo-devo (Specht and Bartlett 2009), ancestral state reconstruction on phylogenies (Sauquet *et al.* 2017), anatomy, genetics, geometric morphometrics, not to mention historical bibliographic synthesis. Often considered as an ancient science, plant morphology is therefore resolutely farsighted. However, we remain convinced that all these conceptual and technical advances allowing a better understanding of the diversity of plant forms will not diminish our wonder at the infinite beauty of these structures... quite the contrary!

7.6. Acknowledgments

We would like to warmly thank Marie-Christine Maurel and Georges Chapouthier for allowing us to participate in this project. We are also indebted to Dr. Alessia Guggisberg (ETH Zurich) for her careful proofreading and pertinent comments, and to Dr. Dario de Franceschi for sending us his book *Terminologie descriptive de botanique* (2016).

7.7. References

Adanson, M. (1763). *Familles des plantes*. Vincent, Paris.

Adler, I., Barabe, D., Jean, R.V. (1997). A history of the study of phyllotaxis. *Annals of Botany*, 80(3), 231–244.

An, C.I., Fukusaki, E.I., Kobayashi, A. (2002). Aspartic proteinases are expressed in pitchers of the carnivorous plant *Nepenthes alata* Blanco. *Planta*, 214(5), 661–667.

Banks, H., Rudall, P.J. (2016). Pollen structure and function in caesalpinioid legumes. *American Journal of Botany*, 103(3), 423–436.

Bauhin, C. (1623). *Pinax theatri botanici*. King, Basel.

Besnard, F., Rozier, F., Vernoux, T. (2014). The AHP6 cytokinin signaling inhibitor mediates an auxin-cytokinin crosstalk that regulates the timing of organ initiation at the shoot apical meristem. *Plant Signaling & Behavior*, 9(6).

Bucksch, A., Atta-Boateng, A., Azihou, A.F., Battogtokh, D., Baumgartner, A., Binder, B.M., Braybrook, S.A., Chang, C., Coneva, V., DeWitt, T.J., Fletcher, A.G., Gehan, M.A., Diaz-Martinez, D.H., Hong, L., Iyer-Pascuzzi, A.S., Klein, L.L., Leiboff, S., Li, M., Lynch, J.P., Maizel, A., Maloof, J.N., Markelz, R.J.C., Martinez, C.C., Miller, L.A., Mio, W., Palubicki, W., Poorter, H., Pradal, C., Price, C.A., Puttonen, E., Reese, J.B., Rellán-Álvarez, R., Spalding, E.P., Sparks, E.E., Topp, C.N., Williams, J.H., Chitwood, D.H. (2017). Morphological plant modeling: unleashing geometric and topological potential within the plant sciences. *Frontiers in Plant Science*, 8(900).

Bock, H. (1546). *Kreüter Buch*. Wendel Rihel, Strasbourg.

Brunfels, O. (1530). *Herbarum vivae eicones*. Ioannes Schott, Strasbourg.

Cesalpino, A. (1583). *De plantis libri*. Giorgio Marescotti, Florence.

Chartier, M., Jabbour, F., Gerber, S., Mitteroecker, P., Sauquet, H., von Balthazar, M., Staedler, Y., Crane, P.R., Schönenberger, J. (2014). The floral morphospace-a modern comparative approach to study angiosperm evolution. *New Phytologist*, 204(4), 841–853.

Chartier, M., Löfstrand, S., von Balthazar, M., Gerber, S., Jabbour, F., Sauquet, H., Schönenberger, J. (2017). How (much) do flowers vary? Unbalanced disparity among flower functional modules and a mosaic pattern of morphospace occupation in the order Ericales. *Proceedings of the Royal Society B: Biological Sciences*, 284(1852).

Chomicki, G., Coiro, M., Renner, S.S. (2017). Evolution and ecology of plant architecture: integrating insights from the fossil record, extant morphology, developmental genetics and phylogenies. *Annals of Botany*, 120(6), 855–891.

Citerne, H., Jabbour, F., Nadot, S., Damerval, C. (2010). The evolution of floral symmetry. *Advances in Botanical Research*, 54, 85–137.

Coen, E., Kennaway, R., Whitewoods, C. (2017). On genes and form. *Development*, 144(23), 4203–4213.

Cronquist, A. (1981). *An Integrated System of Classification of Flowering Plants*. Columbia University Press, New York.

Damerval, C., Jabbour, F., Nadot, S., Citerne, H.L. (2017). Evolution of symmetry in plants. In *Evolutionary Developmental Biology*, Nuño de la Rosa, L., Müller, G.B. (eds). Springer International Publishing, New York.

Damerval, C., Ben Othman, W., Manicacci, D., Jabbour, F. (2018). Distribution area of the two floral morphs of *Nigella damascena* L. (Ranunculaceae): a diachronic study using herbarium specimens collected in France. *Botany Letters*, 165(3/4), 396–403.

Darwin, C. (1859). *On the Origin of Species by Means of Natural Selection*. John Murray, London.

Darwin, C. (1877). *The Different Forms of Flowers on Plants of the Same Species*. John Murray, London.

De Franceschi, D., Octor, H. (2016). *Terminologie descriptive de botanique. Manuel d'initiation à l'analyse du vocabulaire descriptif et glossaire botanique*. De Franceschi, Paimpont.

Della Porta, G.B. (1588). *Phytognomonica*. Orazio Salviani, Naples.

Diggle, P.K. (2014). Modularity and intra-floral integration in metameric organisms: plants are more than the sum of their parts. *Philosophical Transactions of the Royal Society B: Biological Sciences*, 369(1649).

Douady, S. (2011). Phyllotaxis, or how plants do maths when they grow. In *Morphogenesis*, Bourgine, P., Lesne, A. (eds). Springer, Berlin.

Endress, P.K. (1996). *Diversity and Evolutionary Biology of Tropical Flowers*. Cambridge University Press, Cambridge.

Endress, P.K. (1999). Symmetry in flowers: diversity and evolution. *International Journal of Plant Sciences*, 160(S6), S3–S23.

Engler, A. (1898). *Syllabus der Pflanzenfamilien*. Borntraeger, Berlin.

Evans, M., Aubriot, X., Hearn, D., Lanciaux, M., Lavergne, S., Cruaud, C., Lowry, P.P., Haevermans, T. (2014). Insights on the evolution of plant succulence from a remarkable radiation in Madagascar (*Euphorbia*). *Systematic Bio*logy, 63(5), 697–711.

Francé, R.H. (1919). *Die technischen Leistungen der Pflanzen*. Veit & Company, Leipzig.

Fuchs, L. (1543). *New Kreüterbuch*. Michael Isingrin, Basel.

Glidden, J.F. (1874). Improvement in wire-fences. Brevet, U.S. Patent 157,124.

Goethe, J.W.V. (1790). *Versuch die Metamorphose der Pflanzen zu erklaren*. Ettinger, Gotha.

Gómez, J.M., Perfectti, F., Camacho, J.P.M. (2006). Natural selection on *Erysimum mediohispanicum* flower shape: insights into the evolution of zygomorphy. *The American Naturalist*, 168(4), 531–545.

Gould, K.S. (1993). Leaf heteroblasty in *Pseudopanax crassifolius*: functional significance of leaf morphology and anatomy. *Annals of Botany*, 71(1), 61–70.

Green, A.A., Kennaway, J.R., Hanna, A.I., Bangham, J.A., Coen, E. (2010). Genetic control of organ shape and tissue polarity. *PLoS Biology*, 8(11).

Guyenot, E. (1941). *Les Sciences de la vie aux XVIIe et XVIIe siècles : l'évolution de la pensée scientifique, l'idée d'évolution*. Albin Michel, Paris.

Irish, V. (2017). The ABC model of floral development. *Current Biology*, 27(17), R887–R890.

Isely, D. (2002). *One Hundred and One Botanists*. Purdue University Press, Lafayette.

Jabbour, F., Renner, S.S. (2011). *Consolida* and *Aconitella* are an annual clade of *Delphinium* (Ranunculaceae) that diversified in the Mediterranean basin and the Irano-Turanian region. *Taxon*, 60(4), 1029–1040.

Jabbour, F., Ronse De Craene, L.P., Nadot, S., Damerval, C. (2009). Establishment of zygomorphy on an ontogenic spiral and evolution of perianth in the tribe Delphinieae (Ranunculaceae). *Annals of Botany*, 104(5), 809–822.

Jung, J. (1747). *Opuscula botanico-physica*. Otto, Cobourg.

Krishna, S., Keasar, T. (2018). Morphological complexity as a floral signal: From perception by insect pollinators to co-evolutionary implications. *International Journal of Molecular Sciences*, 19(6), 1681.

de Lamarck, J.-B. (1809). *Philosophie zoologique*. Dentu, Paris.

Libiaková, M., Floková, K., Novák, O., Slováková, L.U., Pavlovič, A. (2014). Abundance of cysteine endopeptidase dionain in digestive fluid of Venus flytrap (*Dionaea muscipula* Ellis) is regulated by different stimuli from prey through jasmonates. *PLoS One*, 9(8), e104424.

von Linné, C. (1735). *Systema naturæ, sive regna tria naturæ systematice proposita per classes, ordines, genera et species*. Haak, Lugduni Batavorum.

von Linné, C. (1737). *Genera plantarum*. Conrad Wishoff, Leyde.

von Linné, C. (1753). *Species plantarum*. Laurentii Salvii, Stockholm.

Matamoro-Vidal, A., Raquin, C., Brisset, F., Colas, H., Izac, B., Albert, B., Gouyon, P.H. (2016). Links between morphology and function of the pollen wall: an experimental approach. *Botanical Journal of the Linnean Society*, 180(4), 478–490.

de Mestral, G. (1961). Separable fastening device. US Patent.

Møller, A.P. (1995). Bumblebee preference for symmetrical flowers. *Proceedings of the National Academy of Sciences of the United States of America*, 92(6), 2288–2292.

Mondragón-Palomino, M., Theißen, G. (2009). Why are orchid flowers so diverse? Reduction of evolutionary constraints by paralogues of class B floral homeotic genes. *Annals of Botany*, 104(3), 583–594.

Morton, A.G. (1981). *History of Botanical Science: An Account of the Development of Botany from Ancient Time to the Present Day*. Academic Press, London.

Moyroud, E., Glover, B.J. (2017). The evolution of diverse floral morphologies. *Current Biology*, 27(17), R941–R951.

Moyroud, E., Wenzel, T., Middleton, R., Rudall, P.J., Banks, H., Reed, A., Mellers, G., Killoran, P., Westwood, M.M., Steiner, U., Vignolini, S., Glover, B.J. (2017). Disorder in convergent floral nanostructures enhances signaling to bees. *Nature*, 550, 469–474.

Newell, A.C., Shipman, P.D., Sun, Z. (2008). Phyllotaxis: cooperation and competition between mechanical and biochemical processes. *Journal of Theoretical Biology*, 251(3), 421–439.

Paracelsus (1571). *Astronomia magna, oder die ganze Philosophia Sagax der grossen und kleinen Welt*. Feyerabend, Frankfurt.

Peppe, D.J. *et al.* (2011). Sensitivity of leaf size and shape to climate: global patterns and paleoclimatic applications. *New Phytologist*, 190(3), 724–739.

Ray, J. (1686). *Historia plantarum*. Clark, London.

Rebocho, A.B., Kennaway, J.R., Bangham, J.A., Coen, E. (2017a). Formation and shaping of the *Antirrhinum* flower through modulation of the *CUP* boundary gene. *Current Biology*, 27(17), 2610–2622.

Rebocho, A.B., Southam, P., Kennaway, J.R., Bangham, J.A., Coen, E. (2017b). Generation of shape complexity through tissue conflict resolution. *Elife*, 6.

Reeb, C., Kaandorp, J., Jansson, F., Puillandre, N., Dubuisson, J.Y., Cornette, R., Jabbour, F., Coudert, Y., Patiño, J., Flot, J.F., Vanderpoorten, A. (2018). Quantification of complex modular architecture in plants. *New Phytologist*, 218(2), 859–872.

Ressayre, A., Godelle, B., Raquin, C., Gouyon, P.H. (2002). Aperture pattern ontogeny in angiosperms. *Journal of Experimental Zoology*, 294(2), 122–135.

Rousseau, J.J. (1780–1789). Lettres élémentaires sur la botanique. Collection complète des œuvres, vol. 7(4). Geneva.

Sassi, M., Traas, J. (2015). When biochemistry meets mechanics: a systems view of growth control in plants. *Current Opinion in Plant Biology*, 28, 137–143.

Sauquet, H. et al. (2017). The ancestral flower of angiosperms and its early diversification. *Nature Communications*, 8, 16047.

Sokoloff, D.D., Remizowa, M.V., Bateman, R.M., Rudall, P.J. (2018). Was the ancestral angiosperm flower whorled throughout? *American Journal of Botany*, 105(1), 5–15.

Specht, C.D., Bartlett, M.E. (2009). Flower evolution: the origin and subsequent diversification of the angiosperm flower. *Annual Review of Ecology Evolution and Systematics*, 40, 217–243.

Stebbins Jr, G.L. (1951). Natural selection and the differentiation of angiosperm families. *Evolution*, 5(4), 299–324.

The Angiosperm Phylogeny Group (1998). An ordinal classification for the families of flowering plants. *Annals of the Missouri Botanical Garden*, 85, 531–553.

Theissen, G., Saedler, H. (2001). Plant biology: floral quartets. *Nature*, 409(6819), 469–471.

Théophraste (1988). *Recherches sur les plantes*. Les Belles Lettres, Paris.

de Tournefort, J.P. (1700). *Institutiones rei herbariae*. Typographia Regia, Paris.

Vaillant, S. (1718). *Discours sur la structure des fleurs: leurs différences et l'usage de leurs parties*. Pierre Vander, Leyde.

Wallace, A.R. (1858). On the tendency of varieties to depart indefinitely from the original type. *Proceedings of the Linnean Society of London*, 3, 53–62.

Yan, H., Zhang, H., Wang, Q., Jian, H., Qiu, X., Baudino, S., Just, J., Raymond, O., Gu, L., Wang, J., Bendahmane, M., Tang, K. (2016). The *Rosa chinensis* cv. *Viridiflora* phyllody phenotype is associated with misexpression of flower organ identity genes. *Frontiers in Plant Science*, 7, 996.

Zhong, J., Preston, J.C. (2015). Bridging the gaps: evolution and development of perianth fusion. *New Phytologist*, 208(2), 330–335.

8

Forms of Memory

Robert JAFFARD
INCIA-CNRS,
University of Bordeaux, France

8.1. Introduction

In his book *Die Mneme* published in 1904, Richard Semon (1921) extended the concept of memory beyond the simple recollection of facts or events. Indeed, he considered that heredity also constitutes a form of memory that preserves the effects of experience over generations. He used the term "Mneme" to refer to the fundamental processes underlying both heredity and everyday memory. At birth, each individual already possesses "knowledge" about the external world. This is the memory of the species, also known as "phyletic" or "genetic" memory, which has been shaped over the course of evolution by interactions between organisms and their environment. It allows the individual to communicate and present certain behaviors that are adapted to his external or internal environment. For example, crying in response to cold, pain or a drop in blood sugar levels in infants. It is also the case for certain instinctive motor behaviors such as, for example, the pecking of all small objects in chicks or lurking in the nest at the sight (form) of a predator in seagulls. However, most of the time, the emergence and/or efficiency of this innate knowledge requires a period of maturation and/or some experience. An infant's smile, initially triggered by its own pleasure, will only become a response to a third party's smile after a few months. It is by selectively listening to the call of individuals of their own species that the chicks memorize – form a neural representation or "pattern" – of this call, long before they themselves are able to express it vocally. It is also through progressive trial and error that the chick will end up only pecking seeds, and it'll do this with great precision. Thus, if the phyletic

The Explosion of Life Forms,
coordinated by Georges CHAPOUTHIER and Marie-Christine MAUREL. © ISTE Ltd 2020.

memory contains *innate capacities* to recognize and react to particular *stimuli*, via the architecture of the primary sensory and motor cortex, it quickly becomes dependent on new acquisitions, which complete it and constitute the beginning of the individual memory. This memory ranges from habituation, which is the most elementary form of "behavioral plasticity", to episodic memory, which is the ability to remember events experienced personally.

8.2. The polymorphism of memory

In the common meaning of the word, memory is the ability to store information so that we can later reproduce it or use it to guide or modify our present behavior. Such a definition does not capture the complexity of the phenomenon, which we know covers a wide variety of situations: I know how to play chess and ride a bicycle, I know the capital of Egypt and the date of Montaigne's death, I remember the day my son was born and the last time I met my cousin Pierre, but I also remember – or know – that I have to walk my dog every night and that I have to attend a meeting tomorrow at 10:00 a.m. These different examples, far from exhausting all the situations (for example, I can also perform mental arithmetic operations) illustrate this complexity.

The idea that memory is not a unitary ability has a long history. In *The Principles of Psychology* (1890), William James wrote two separate chapters, one on habits and the other on memory. In the latter, he distinguishes primary (short-term) memory from proper or secondary (long-term) memory, in which he further compares "memory", which "refers to *my* past", with "knowledge". In 1799, Maine de Biran described three forms of memory: representative memory, mechanical memory and sensitive or affective memory, the latter two being "empty of ideas or representations". Bergson's (1896) conceptions are very similar in his distinction between habitual memory, "which plays on our past more than it represents it", and pure memory, which "allows us to evoke the past in the form of images".

It is essentially during the last 40 years that psychobiological investigations conducted in animals and humans have made it possible to go beyond philosophical discourse and intuitions, by showing that the way the brain stores the information it receives reflects this polymorphism.

It can be considered that the decisive discovery was made in 1962 during studies conducted on the patient Henri Molaison (HM) who, having become deeply amnesic following bilateral resection of his medial temporal lobes (MTL), proved to be perfectly capable of learning and retaining perceptual-motor skills without having the slightest *memory* of having practiced the exercise before (Milner 1962). It later became clear that a wide range of learning and memory (dissociation) skills were retained intact in amnesic patients, as well as in animals (mammals) with similar brain lesions. Conversely, other brain injuries, such as neostriatum, were found to have opposite effects (double dissociations). All these observations led to the organization of memory into systems, each of these memory systems being characterized both by its ability to process (store, retain and use) a particular type of information and by the brain modules that perform this processing (see Figure 8.1).

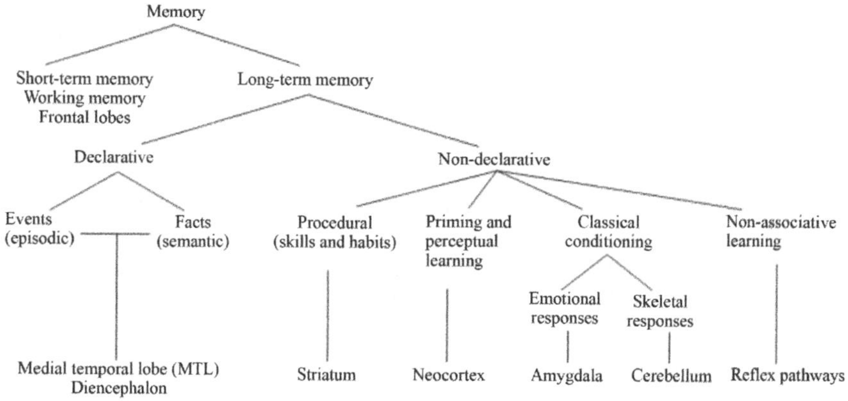

Figure 8.1. *Mammalian memory organization. The brain structures mentioned are considered particularly important for the form of memory considered. It is modified according to Squire and Dede (2015) (see Jaffard 2011). The order in which these memories are presented in the text is not exactly the same order (from right to left) as in the figure, as it introduces other taxonomic criteria*

8.3. Non-associative memories

Non-associative memories are formed as a result of the simple *presentation of a stimulus.*

8.3.1. *Habituation and sensitization*

8.3.1.1. *Habituation*

It is the progressive decrease in the reflex response to a *stimulus* as it is repeated that allows us to learn that a *stimulus* is (or has become) "unimportant". Generally speaking, habituation is of short duration.

It is the most basic form of behavioral plasticity. It exists at all levels of the phylogenetic tree, from protozoans to insects and mammals.

For example, the *acoustic startle reflex* in rodents is a defensive reflex that occurs in response to an unexpectedly loud sound, but gradually diminishes with repetition. The behavioral *orienting response* (OR) to a new stimulus (e.g. visual stimulus) is an innate attentive response to novelty that disappears with the repetition of the stimulus (see below). A major characteristic of habituation is that it is specific to the stimulus that caused it, which distinguishes habituation from other forms of decremental responses such as fatigue. For example, the habituation of the *Nereis* worm's contraction reflex to a mechanical shock is independent of its habituation to a moving shadow which, presented immediately afterwards and repeated, produces a reflex response that disappears with the same time course as the previous one (Clark 1960). Experiments in rats show that ORs initially directed to either of two composite stimuli (sound + continuous light and click + flash light) disappear (habituation), but reappear if the composition of these stimuli changes. For example, the "sound + flash light" presentation will result in the reappearance of the OR. This "dishabituation" is nevertheless suppressed by *lesions in the hippocampus* (Honey *et al.* 1998), which shows that this structure may (also) be necessary for the full expression of the most basic form of memory – here by constructing an accurate representation of the relationships between *stimuli* (a high-level *relational memory*, see section 8.7.1.2) that "controls" the habituation of the OR. This is obviously not the case in Aplysia, where the habituation of the defensive gill withdrawal reflex is only controlled by about 40 sensory neurons, which innervate the skin of the siphon, and by six motor neurons, which innervate the gill.

8.3.1.2. *Sensitization*

It is, in contrast to habituation, the increase in reactivity to a stimulus following exposure to another stimulus that is most often intense, painful or threatening. For example, the threat of a predator results in a decrease of the trigger threshold for protective behavior, which can also be caused by other stimuli (in other words, unlike habituation, sensitization is not specific to a stimulus).

Experimentally, the *acoustic startle reflex* is potentiated by the prior administration of an electric shock or by any associated stimulus of normal intensity that causes fear (in other words, by classical conditioning, see section 8.4.1). In humans, sensitization can be studied under the same conditions, but using a "correlate" of the startle response, the galvanic skin response (GSR) or skin conductance response (SCR), which is an index of the emotional state of fear, anxiety or surprise.

Lastly, exploration of an environment that is in itself a learning experience (described as "latent" by Tolman, 1948) decreases (habituation) with exploration, but can be reactivated (dishabituation or sensitization) if the environment contains new elements (an object, for example).

8.3.2. *Priming*

Sensitization increases the likelihood of response to a stimulus. The same is true for priming, which increases the ability of humans to identify an item through prior exposure to that item or related items.

In the 1970s, Elizabeth Warrington and Lawrence Weiskrantz (1968) showed that *amnesic patients* were able to "retrieve" words (e.g. "porch") or images (e.g. an airplane) that had been presented to them provided that, in a subsequent test, they were aided by the presentation of these same items in a highly fragmented (not directly identifiable) form.

This result seemed to show that amnesia is linked to a deficit in *recovery* or *restitution*, in other words, a lack of access to information stored elsewhere and *not to the absence of storage*. It was later shown that, in this type of situation, the apparently normal performance of amnesic patients strictly depends on the instructions given during the test. Only an *implicit* or *indirect* instruction is effective: "From what I show you [fragmented item], tell me what *immediately comes to mind.*" On the other hand, the *explicit* or *direct* instruction: "Based on ..., try to find the word [or image] that was *in the list* that was presented to you earlier" is totally ineffective because it "directs" the patient to the explicit memory of those items that he or she does not have.

Priming corresponds to a form of memory known as "implicit" (or unconscious) memory, preserved in the amnesic syndrome and *distinct* from the *explicit* (or conscious) memory, the impairment of which characterizes this syndrome. In normal individuals, these two forms of memory independently underlie the memory of the same information (words, images, sounds, etc.). There are two forms of priming: perceptual priming based on form and conceptual priming based on meaning. Let's take the simple example of a list of about 10 words in which the word mango and the word birch appear. Whether it is a free recall ("Give the words from the list learned."), a cued recall ("What was the word on the list beginning with man?"), or a recognition test ("Was this word on the list – yes or no?"), amnesic patients fail these *direct tests*. However, they do pass *indirect tests*: "Give the first word that comes to mind" after the presentation of *man* (answer: "mango"), or "Give the first name of a tree that comes to mind" (answer: "birch"). These two forms of priming can be dissociated. For example, *Alzheimer's disease* spares perceptual priming that relies on the primary sensory cortices that resist the disease, but impairs conceptual priming that involves the frontal and temporal lobes which are damaged (Fleischman *et al.* 2005).

Phenomena similar to human perceptual priming have been reported in some birds.

8.3.3. *Perceptual learning*

Perceptual learning (Gilbert *et al.* 2001) is conceptually similar to priming in that prior exposure to a stimulus facilitates its identification (perceptual memory). It is different from priming in that it is only very gradually acquired (hundreds of exposures are required) and is extremely specific to what has been learned (not generalizable or transferable to other, even related, situations). It reflects the continuous assimilation of the sensory environment throughout life, the way we perceive the world, and also explains expertise, whether it be a specialist doctor who interprets an X-ray or a farmer who instantly identifies the sex of a chick. Perceptual learning is measurable through the progressive improvement of performance during sensory discrimination tests. In the *visual domain*, these tests consist of identifying increasingly small differences in visual attributes such as position, texture or form of stimuli. In other sensory domains, the aim is to improve pitch discrimination or tactile perception of changes in vibration frequency.

Like priming, perceptual learning is implicit. It progresses without the need to reward "right answers". Moreover, in addition to being highly specific to the type of discrimination and the context in which the training was carried out, the memory of perceptual learning is retained for a very long time. For example, in a "texture discrimination" experiment (visual detection of a series of three lines whose orientation is different from the set of hundreds of others), the level of performance achieved after four or five days can be maintained for at least three years without the need for further training (Karni and Sagi 1993).

Lastly, numerous experimental observations suggest that this type of learning relies on changes in the properties of neurons that are involved in the early stages of cerebral processing of sensory information (e.g. V1 areas of the visual cortex), which would "tune" their responses to the specific sensory training attribute.

8.4. Classical conditioning

My heartbeat, which quickens as I cross the crossroads where this terrible accident took place; your irrepressible urge to smoke a cigarette as soon as you smell coffee; her aversion to chocolate éclairs, which she was eating greedily the day she fell ill, etc.

All these behaviors are the result of classical conditioning.

8.4.1. *Operational definition, rules and varieties of classical conditioning*

Classical conditioning is conditioning in which an organism learns that an *S1 stimulus* predicts the occurrence of another *S2 stimulus*, where S1 is called the "conditional stimulus" (CS) and S2 is called the "unconditional stimulus" (US). The US is a significant event (positive or negative) for the organism under consideration.

Virtually unknown to Westerners until the late 1920s, salivation conditioning was first discovered and studied by Ivan Pavlov in the 1890s. He demonstrated that various sound or visual stimuli (S1) presented a few seconds before food (S2) acquired the attribute of provoking a conditioned response (CR) of salivation (which he called "psychic").

Currently, *motor conditioning of the palpebral reflex* (CPR) (closing of the eyelids) in response to a sound (CS) previously associated with a pulse of air directed at the cornea (US) and *fear conditioning* (FC) produced by the association of a sound or visual stimulus (CS) with an aversive stimulus (electric shock, intense noise) are well studied in humans and rodents, where the brain structures (cerebellum and amygdala, respectively) and the cellular mechanisms underlying them are remarkably similar.

Taste aversion conditioning (CTA), discovered in the 1950s in rats by John Garcia *et al.* (1955) and virtually present in all mammals, is very different from other conditioning, as the interval between CS and US is several hours (20–30 seconds and less than 1 second are the maximum possible for CPR and FC respectively) and a single association is sufficient. Furthermore, while CPR and FC require the integrity of the hippocampus as soon as the CS and US are separated, CTA operates without the hippocampus and relies on the insular cortex.

However, in addition to the diversity of its manifestations and its cerebral supports in mammals, there is also a diversity of organisms that are equipped with it. Classical conditioning emerged very early in evolution. Present and much studied in insects such as fruit flies and bees, it has even been demonstrated in sea anemones, indicating that it can be driven by simple neural networks.

Nevertheless, it was the work on mammals from the 1960s onwards that made it possible to forge the current theory of conditioning.

8.4.2. *Contemporary theory of classical conditioning*

8.4.2.1. *From contiguity to contingency and associability of stimuli*

In an article entitled "Pavlovian Conditioning: It's Not What You Think It Is" published in 1988, Robert Rescorla wrote: "Classical conditioning is not a stupid process by which the organism willy-nilly forms associations between any two stimuli that happen to co-occur. Rather, the organism is better seen as an information seeker, using logical and perceptual relations among events, along with its own preconceptions, to form a sophisticated representation of its world."

What results did he rely on to support these assertions?

Since his first work in 1966, Rescorla had shown that the rule of *contiguity*, hitherto the cornerstone of classical conditioning, was neither sufficient nor even necessary, and that it had to be replaced by the concept of contingency with two consequences. First, the strength of a conditioning depended on the probability that CS would be followed by US, in other words, on the "reliability" of the prediction. Second, for a *null contingency* (absence of any contiguity), the subject was learning. He learned that CS "guarantees" the absence of US.

As we have seen, the contiguity rule is also refuted by the existence of taste aversion conditioning (CTA), where the perception of the taste of a food (CS) and visceral discomfort (US) are separated by several hours. Work on CTA has also shown that associations are selective for certain stimuli. For example, food is easily associated with discomfort by its taste, whereas it is not associated with visual and/or auditory stimuli presented at the time of consumption. This observation and many others contradict what Pavlov thought: "Any natural phenomenon chosen at will can become a CS... any visual, sound, or odor stimulus, as well as the stimulation of any part of the skin ..." (Pavlov 1928, p. 58). In fact, and this is the result of evolution, animals learn little or nothing about improbable associations in their environment.

However, it was Léon Kamin's highlighting of the blocking effect in the late 1960s that was the decisive factor which led to abandoning the old rules. In his *princeps* experiment, Kamin showed that if the pairing of two CSs were presented simultaneously (CS1: sound and CS2: light) to an electric shock (US), both provoke a conditioned emotional response (CER), the prior association of one of them (for example, CS1) leads to a blocking of the other association (CS2-US) (Rescorla 1988a, pp. 340–41). In other words, if you are allergic to eating chicken, you will probably not associate your allergic reaction to potatoes that you have just eaten with your chicken.

8.4.2.2. *Contingency, "surprise", prediction error and dopamine*

Another way to explain the blocking effect is to consider that the organism is not surprised by the occurrence of the electric shock that has already been entirely predicted by the presentation of the sound (or the allergy already predicted by the consumption of chicken!). This perspective has been formalized by Rescorla and Wagner (1972) in a simple equation that allows us to calculate the speed of a conditioning and to account, among other phenomena, for the blocking effect. This equation indicates that the increase in the strength of association ΔV of a CS with a US is $\Delta V = A\ (\lambda\text{-}V)$. The important factor in this equation is that the conditioning depends on the difference between the initial association force V, which, for the subject, corresponds to the probability of occurrence of the US, and the maximum strength of association λ that can be developed with the US used. In the case of blocking, since the strength of association of CS1 with the US is already at its maximum ($\lambda\text{-}V = 0$), CS2 can no longer be associated. In other words, there is an increase in the strength of association – or there is conditioning – only if the subject is "surprised" by the difference between the two, that is to say, if $\lambda - V > 0$. This difference is also called "prediction error". In mammals, many results show that this prediction error (surprise) is coded by the activity of dopaminergic neurons, which reinforces the strength of association (Schultz 2016).

8.4.3. *The importance of classical conditioning*

By anticipating the occurrence of a significant positive or negative event, classical conditioning allows many organizations to respond more effectively to this event. The anticipated response of salivation improves the assimilation of food, that of eyelid closure avoids damaging the cornea, while fear conditioning anticipates a defensive reaction to the warning of danger. In this respect, conditioned taste aversion is particularly noteworthy, in that an association between taste and nausea is established at once and leads to a lasting avoidance of the consumption of a potentially lethal food. However, classical conditioning can also have more negative consequences. For example, all environmental stimuli (CSs) associated with the consumption of tobacco, alcohol, heroin or cocaine (USs) are powerful inducers of the corresponding craving (CR). In addition, a homeostatic "compensatory conditioned response" develops when the same substance is administered repeatedly in the same environment, leading to an attenuation of its effects. For some, including Shepard Siegel (2016), most accidents attributed to a heroin "overdose" could be linked to taking the substance in a new environment, thus leading to the suppression or weakening of the usual compensatory conditioned response.

8.5. Instrumental conditioning

In the classical conditioning discovered by Pavlov, the subject learns "passively" to associate stimuli from his environment with one another. In instrumental conditioning, which Thorndike – who was unaware of Pavlov's work at the time – systematically studied around the same time, the subject learns the consequences of his own responses in this environment. Thus, when a rat is placed in a box with a lever to operate a food dispenser, explaining why it salivates is a classical conditioning theory, whereas explaining why it presses the lever is an instrumental conditioning theory. In this type of conditioning, the important event is of the same nature as the US of classical conditioning: it is a reinforcement (Rf).

The Rf that follows a response can be a reward, such as food, or a punishment, such as an electric shock, which respectively trains to increase and decrease this initially spontaneous response. This description is "neutral": it reports observable behavior without reference to the "mind". This is the position of the *behaviorism school*. The terms "response" could be replaced by "voluntary action" and "entails" could be replaced by "with the aim of obtaining" and invoke knowledge or a representation of the situation. The reference to such mental processes characterizes the *cognitive approach* to instrumental conditioning which, as we shall see, essentially concerns mammals.

Instrumental conditioning exists in a wide range of species, from mollusks (including aplysia) to insects (bees and fruit flies) and other arthropods.

8.5.1. *Law of effect, stimulus-response (S-R) theory and "habits"*

For Thorndike (1911), conditioning results from an increase (or decrease) in the connection between the nerve centers processing one or more S stimuli present at the time of the response (or the set of ΣS stimuli representing the situation) and those controlling that R response. Thorndike's Law of Effect thus states that, of all possible responses, those followed by a satisfactory state (reward) will be strongly "connected" to the situation (ΣS) or to a particular S stimulus. Conversely, responses that are followed by an unpleasant state (punishment) will be "disconnected" from that situation.

This stimulus-response (S-R or ΣS-R) theory of instrumental conditioning refers to habits, "routines" or procedures that are gradually acquired and performed automatically and unconsciously (implicit procedural memory, see Figure 8.1).

However, instrumental conditioning is not limited to this aspect alone.

8.5.2. *From S-R theory to cognitive theories*

An alternative to the S-R theory is that R response is not automatically *triggered* by S or ΣS, but is controlled by the subject's knowledge of the consequences, C, of his R response. In other words, as Tolman assumed as early as 1932, this response can be intentional and can be qualified as an action; it is directed towards a goal, of which the subject has a representation. This hypothesis is confirmed by the following experiments.

8.5.2.1. *Instrumental conditioning and encoding of the relations between the R response and its C consequences*

These experiments show that if, after conditioning, the reward used as reinforcement Rf loses its initial motivational value or relevance to the animal, the response may be significantly diminished or even eliminated. Two methods are used: devaluation of the reinforcement and degradation of the instrumental contingency.

The devaluation of the reinforcement Rf consists of either letting the animal consume the food used as Rf in sufficient quantity just before the test to create a *specific satiety*, or in provoking a taste aversion conditioned to this food (see classical conditioning) by following its consumption with a visceral discomfort induced by the administration of a toxic substance such as lithium chloride.

Contingency degradation is based on the hypothesis that the animal learns that a certain response leads to a certain reinforcement; for example, pushing a bar to the left results in an aniseed-flavored pellet. Therefore, if this reinforcement is delivered independently of this response or action, the animal "learns" that there is no longer a relationship between the two, and this action is therefore "useless".

The use of these two methods clearly demonstrated the existence of R-C associations and thus, by extension, S-R-C associations. For example, in Colwill and Rescorla's *princeps* experiment (1985), rats were first taught, in the same apparatus, to perform an R1 response (pressing a lever) to obtain the Rf1 reinforcement (food pellets) and then a different R2 response (pulling a chain) to obtain another Rf2 reinforcement (sucrose solution). Since one of the two reinforcements was then devalued (e.g. Rf1) by conditioned taste aversion, the authors observed that the response leading to the devalued reinforcement (R1) decreased very strongly relative to the other (and vice versa). Similar results were observed in monkeys and humans. For example, once monkeys have learned to associate different images with two types of food, one of the two foods becomes less "desirable" because of its recent consumption (specific satiety), so the monkeys choose the images associated with the other food.

8.5.2.2. *Removal or reduction of the contribution of R-C to the benefit of S-R (automation) through overtraining*

Numerous results show that devaluation, which is very effective after limited training, becomes ineffective after overtraining. For example, in the previous situation in rats, the number of R1 and R2 responses recorded in the test following the devaluation of either reinforcement tends to become the same as the duration of training increases. In other words, habit (S-R) becomes predominant and the value of the Rf is "transferred to the lever".

8.5.3. *The two facets of instrumental conditioning*

Instrumental conditioning is based on two systems: a cognitive system, based on the knowledge of a relationship between a response (action) and its consequences (R-C), and an automatic S-R system, also called "habit". With practice, the same response is less and less controlled by the cognitive system and therefore loses its flexibility (e.g. changing one's usual daily S-R route to go to the baker's).

This dual system has been frequently referred to by some authors, including William James (1890, Chap. 4, p. 6), who pointed out that "habits diminish the conscious attention with which our actions are performed", and Edwin Schrödinger (1990, p. 158), who considered that "in a metaphorical way, consciousness is the tutor who supervises the education of the living substance, while leaving his pupil alone to perform all the tasks for which he is already sufficiently trained". It was also Edward Tolman (e.g. 1948, pp. 206–207) who noted that rats in a maze find it difficult to abandon the path they have repeated many, many times when given the opportunity to follow a shortcut, whereas rats with limited training quickly adopt it.

The neural supports of these systems are relatively well known in rodents and primates (Balleine and O'Doherty 2010). Schematically, they are based on two *corticostriatal circuits*, one of which is "associative" and underpins R-C, and the other, "sensorimotor", underpins S-R. In rodents, it is possible, through lesions targeted on these circuits, to obtain an instrumental conditioning that *exclusively* relies on S-R (insensitive to devaluation) or R-C (which remains sensitive to it, even after overtraining). Stress induces the hypofunctioning of the "associative" circuit and therefore favors S-R to the detriment of R-C (automation). In addition, updating the "value" of a reinforcement involves the amygdala (which reassesses) and the insular cortex (which "stores" the new value).

These observations are important for studies that dissociate learning strategies based on habits *stricto sensu* (procedural memory) from those dependent on the hippocampus (spatial memory, for example).

8.6. Procedural memory as a "memory system"

In mammals, procedural memory is a heterogeneous set of learning and retention abilities. These abilities have commonalities, such as 1) being spared in the human amnesic syndrome or after damage to the hippocampus in animals; 2) being, conversely (double dissociation), often deteriorated by dysfunctions or damage to the dorsal striatum (Parkinson's and Huntington's diseases); 3) being acquired progressively and in an unconscious or implicit way in humans. Upon these criteria, procedural memory includes habits (the automated S-R component of instrumental conditioning, see section 8.5) and, in humans, skills.

8.6.1. *Habits: double functional dissociations in mammals*

8.6.1.1. *In rodents*

Different tests have been designed so that, in the same situation, learning can be achieved using two strategies: a habitual S-R (or procedural) strategy and a spatial strategy, based on a representation of space (spatial or place learning). There are essentially three devices: the *T-shaped maze* (Tmaze) and the *radial maze* (Rmaze), in which animals learn to find food, and the *Morris water maze* (MWM) where, by swimming in opaque water, they must learn to find a small invisible (submerged) platform that allows them to escape the water. In these devices, the habitual S-R strategy may be a motor "response" strategy based on an orienting body movement (e.g. to the right or left), when a choice is presented (Tmaze), or a "guiding" strategy of using *internal cues* to the maze, such as arm lighting (Rmaze) or an object placed above the platform (MWM). In all these situations, *external cues* to the device also allow the use of a *spatial strategy*.

The existence of two independent (separable) memory systems can be illustrated by the following results (White *et al.* 2013):

a) when spatial and habitual S-R strategies are also relevant to learning the task, animals with hippocampal lesions all use the S-R strategy (e.g. guiding in the MWM), while animals with dorsal striatum lesions all use the spatial strategy;

b) in the Tmaze where both strategies are possible, the same rats first use a spatial strategy, which, with overtraining, is gradually replaced by an S-R response strategy, a change in strategy that is akin to automation. At this stage, a return to the spatial strategy is still possible (in other words, the spatial memory has not been "erased", its use is "masked");

c) after five days of training, mice forced to either use the S-R guidance strategy (Group A) or the spatial strategy (Group B) in the MWM showed an increase in the volume of the striatum but not the hippocampus (Group A), and the hippocampus but not the striatum (Group B), measured by a high-resolution MRI (Lerch *et al.* 2011).

Conceptually similar results have been reported in humans.

8.6.1.2. *In humans*

Placed in a virtual radial maze where both strategies can be used, the natural propensity to use the spatial strategy is associated with activation (functional MRI) and a higher grey matter density in the hippocampus than in the striatum (caudate nucleus), and vice versa for the use of a procedural S-R "response" strategy. Just like in rodents, an initially spatial strategy evolves into a procedural strategy (e.g. (Bohbot *et al.* 2007)).

These results and those of Lerch *et al.* (2011) refer to the observations of Eleonore Maguire and her team who, in 2000, reported an increase in the volume of the posterior hippocampus of London taxi drivers, the cause of which was indeed their professional practice and its duration. In an experiment carried out by the same team, volunteers traveled the streets of a virtual city by either playing a "taxi driver" role, which involves imagining the shortest routes to get from one point to another (spatial strategy), or the "bus driver" role, which involves repeating the same route (S-R strategy). In the first case, the activation of the (anterior) hippocampus was greater than that of the striatum (head of the caudate nucleus), and vice versa in the second case (Hartley *et al.* 2003).

8.6.2. *Human procedural memory and its cerebral supports*

8.6.2.1. *Habits*

Concurrent discrimination of objects: a "habit" in monkeys and "declarative" in humans

In monkeys, the "concurrent" discrimination test is frequently used, in which several pairs of objects – often eight – are presented each day one after the other in pseudo-random order, and the same number of times. In each pair, the choice of an object, always the same and regardless of its position, makes it possible to obtain a reward. However, it takes the monkey several hundred trials to master the task, whereas human subjects do it in a few or dozens of trials. This difference – which is not due to the possibility of naming the objects used – is related to a difference in strategy: to learn this task, humans use their declarative system based on the hippocampus, whereas monkeys use their procedural incremental S-R system based on the striatum. Patients with amnesia of hippocampal origin only acquire discrimination very gradually, in fact much like a normal monkey (or one with hippocampal lesions). As we have already noted several times, these patients have no memory of their learning, which is only expressed through their "performance" (Bayley *et al.* 2005).

Measuring human habits by "predicting the weather"

Humans spontaneously use their declarative memory to learn what non-human primates master with their procedural memory. To "force" humans to use their procedural S-R memory, researchers use tests in which the S-R relationship is "probabilistic". The most widely used test is the weather prediction task. It is presented as a game in which one must predict the weather – rain or sunshine – from tarot cards that are presented side by side and that can form several configurations. Each configuration only gives a *probability* ($p < 1$) and *not a certainty* ($p = 1$) on the answer to give. Thus, during the learning process, the subject must, for each presentation, give his answer ("rain" or "sunshine"), which is immediately confirmed or invalidated by the experimenter (feedback). For example, if, for a given configuration, the probability of "sunshine" is 3 out of 4 (0.75), the answer "sunshine" will be confirmed 3 times out of 4 and invalidated 1 time out of 4. With practice, normal subjects gradually improve their ability to correctly "predict" the weather. They go from giving random answers (50%) to giving 70–75% correct answers in about 50 trials. Patients with amnesia show the same improvement, but again, unlike normal subjects, they do not remember the test. Conversely, Parkinson's patients who remember the test do not learn it. Lastly, neuroimaging (functional MRI) studies carried out on healthy volunteers show activation of the hippocampus at the beginning of learning (probably related to the attempt to use a declarative strategy) followed by its *deactivation* associated with activation of the caudate nucleus, as performance based on probabilistic S-R acquisition improves (Poldrack *et al.* 2001).

8.6.2.2. *Skills: functional dissociations*

Skills represent what we have learned about how we "do" things in everyday life; they can be motor, perceptual or cognitive. For example, learning Morse code involves both motor and perceptual-motor skills; understanding ("decoding") a message is a perceptual skill.

In addition to the perceptual-motor skill involved in the mirror tracing task, which revealed that the HM patient could "learn" (see section 8.2), maintaining contact with a rotating target (rotary pursuit task) is also a skill preserved in amnesia. The same is true for perceptual skills such as learning to read mirror words (mirror reading task) and motor skills such as those assessed in the serial reaction time task (SRT). In this test, the subject must respond as quickly as possible to the appearance of visual stimuli – four in number – which, aligned horizontally, appear successively on a screen. To do this, the subject must press the key on a keyboard that corresponds to the position of the stimulus on the screen. The important thing in this test is that the SRT decreases over repeated sequences without the subject having to remember these repetitions. Cognitive skills are assessed by the ability to solve certain problems, such as the Tower of Hanoi problem where, placed in front of a support with three vertical axes, the subject learns by trial and error to move the N discs of decreasing size that are stacked around the left axis, one by one, in order to bring them into the same position around the right axis, using the central axis. The minimum number of movements required is 2^N-1 (e.g. 31 for 5 discs). Largely preserved in the amnesic syndrome, these skills involve the dorsal striatum (Squire and Kandel 2015).

8.7. Declarative memory

As noted above (see section 8.2), William James (1890) distinguished between memory that "must be dated in *my* past" and memory that "failed to come up with any one particular setting". He added that "to ask a man whether he *remembers* that the sun shines [...] would be a misuse of language".

8.7.1. *Episodic and semantic memory: definitions, properties and relationships*

8.7.1.1. *Definitions and properties*

Episodic memory: "I remember"

The term "episodic memory" was coined by Endel Tulving in 1972 to describe a type of human memory that can be applied to the autobiographical memory of events. In order to be experienced as a memory, the event (what) recalled must be mentally represented in its particular temporal (when) and spatial (where) context – and with a sense of "having been there". This presupposes a mental time travel in the past, accompanied by a particular state of consciousness, which is called autonoetic consciousness.

"True" recollection is different from the simple sense of familiarity we may experience, for example, when meeting someone for the second time. In the first case, it is mentally placed in the "context" (where and when) of the first encounter, whereas in the second, it is not (I know that...). Most specialists consider that the feeling of familiarity (the what) can be based on the parahippocampal region alone and that the hippocampus is the only thing that allows one to establish its binding with the context to make it a recollection (Jaffard 2011).

Semantic memory: "I know that"

It is the totality of our knowledge about the world represented in an organized form: establishing *relationships*, *groupings*, *categorization*, etc. Used in the *present*, this memory does not imply a mental journey into the past, *nor* does it refer to a particular spatial-temporal context. You know that the Eiffel Tower is in Paris without necessarily remembering when you learned about it. This knowledge is based on a noetic consciousness (does not refer to oneself), the subject now being an "observer" and no longer an "actor" like in episodic memory.

8.7.1.2. *Hierarchy and common properties*

Semantic memory is *not always* the result of memories which, repeated in different contexts and thus losing their spatial-temporal markers, would become, by *decontextualization*, knowledge. For Endel Tulving, information must first be encoded in semantic memory before being encoded in episodic memory (Jaffard 2011). In fact, adult patients suffering from *developmental amnesia* linked to selective lesions of the hippocampus around their birth almost have normal semantic memory, despite a very deficient episodic memory (Vargha-Khadem *et al.* 1997).

Episodic memory and semantic memory have flexibility in common. The account of the same (episodic) event can vary in the choice of words, the details used to describe it, and allusions to other similar events. The expression of our knowledge (semantics) is also modified by conditions or "demand" (e.g. MCQ or dissertation or application). This flexible expression involves representations that relate (create links) between information recorded at different times on the basis of their common elements. This is one of the reasons why declarative memory is often referred to as relational memory by "default", a memory whose cardinal adaptive property is the capacity for inference (deduction) that exists in most mammals (see, for example, Zeithamova *et al.* 2012).

8.7.1.3. *Episodic memory, semantic memory, imagination and false memories*

Remembering the past and using knowledge to imagine the future

Over the past 15 years or so, it has been established that memory, in the traditional sense of the term, in other words, memory from past (retrospective) experiences, and what is known as "memory of the future", are, both in terms of their behavioral expression and the brain mechanisms underlying them, two sides of the same process.

Put another way, "mental journeys" can not only be made to the past but also to the future (Mnemosyne has both faces of Janus). The quality of a subject's episodic memory is correlated to his ability to construct – to imagine – coherent, detailed and personal future scenarios. Amnesic patients only provide impoverished, impersonal constructions of commonplaces and generalities. Neuroimaging studies lead to the same conclusions. The hippocampus and other activated brain structures are largely the same when a subject mentally evokes an event from his own past (episodic autobiographical memory) or when he imagines a scene in his own future. In this case, his knowledge (semantic memory) guides the mental construction of these "future *scenarios*" (see, for example, Schacter *et al.* 2017).

Episodic memory, semantic memory and false memories

Being sure that you remember an event that never happened is a false memory. For Elizabeth Loftus (2003, p. 872), "people's memories are not only the sum of what they have done...[they] are also the sum of what they have thought, what they have been told, what they believe". Many experiments have shown that the formation of these false memories can be based on imagination (see above) and/or suggestibility, on retrospective distortions with the integration of *erroneous* information into a pre-existing memory, and finally on extrapolation, through a process of semantic generalization.

The DRM test (for Deese, Roediger and McDermott, its developers) illustrates this process of generalization. It consists of presenting a list of words that are all linked to a theme word that is not on the list and will serve as a "critical lure". For example, butter, toast, crust, rye, etc. will be the words on the list that are actually presented and "bread" will be the non-presented "critical lure". Experiments show that 40–70% of subjects identify the word "bread" as a word on the list. This type of false memory reflects an "incursion" of semantic memory into episodic memory. It is, in a way, like the "price to pay" for the structuring nature of semantic memory, which organizes our knowledge to optimize its use. According to this hypothesis, a dysfunction of the anterior lateral temporal lobe (ATL), the storage site of semantic memory, *reduces* false memories in the DRM test without affecting "real" memories. Conversely, false memories increase in the elderly, where the weakening of "true" episodic memory resulting from a hypofunctioning of the hippocampus appears to be compensated by a hyperfunctioning of the ATL, reflecting a greater use of semantic memory (Straube 2012).

8.7.2. *Episodic memory in animals?*

Answering this question affirmatively would mean that some of them are able to integrate the content of an event (what) into its context (where and when), which is the case. But if one also requires that this memory involves the two subjective criteria of "mental time travel" and "autonoetic consciousness", it is clear that this question cannot be answered. For this reason, animal memories that meet the what-where-when (WWW) criterion are termed episodic-like memory.

In a now famous experiment, Nicola Clayton and Anthony Dickinson (1998) used the natural propensity of the western scrub jay to hide various types of food (worms, insects, seeds, etc.) in a wide variety of locations in order to build up reserves. Their laboratory experiments show that these birds remember *where* they hid such and such food (*what*) – which was already known – but in particular, they remember how long ago they hid it (*when*).

Subsequent experiments show that, in different situations, rats also develop a WWW memory. This finding has been extended to other species, including cuttlefish and zebrafish (Hamilton *et al.* 2016). However, the demonstration of WWW memory in monkeys has proven to be much more problematic than previously thought.

All these results raise many problems, particularly in terms of evolution (see for example (Murray *et al.* 2018, p. 408)), and should lead to considering the issue from another angle. Such an example is determining whether certain properties of human episodic memory are shared by other species and whether they are based on common *neurobiological mechanisms*. Temporal order memory ("flow of events"), which structures episodes, is probably one of these shared properties (Eichenbaum 2013).

8.8. Short-term memory and working memory

8.8.1. *General characteristics*

Short-term memory (STM) allows many living things to temporarily hold a limited amount of information in memory. In our daily lives, this means remembering the telephone number that was just read until it was dialed on a keyboard. The memory span refers to the number of elements that can be retained in this way. It averages 7 for a sequence of numbers and 5 for a sequence of spatial positions, but depends on many factors including the type of test, the nature, novelty and complexity of the material to be retained, as well as the age and training of the subject.

Working memory (WM) allows the manipulation of the information maintained in STM, in order to modify it for a given purpose. For example, a series of numbers can be ordered according to a particular rule, words can be sorted alphabetically, objects can be sorted by increasing size, etc. Information may be removed and/or replaced by other information (update, see below). More broadly, the WM is of major importance in many cognitive skills such as mental arithmetic, reasoning, language comprehension, planning, decision making or the formation of "mental maps" of space.

STM and WM interact with long-term memory (LTM). Firstly, because all information derived from the LTM passes – just like information perceived in the environment – into the STM/WM. Secondly, because the manipulation or organization of information in WM involves the use of rules, knowledge or memories stored in the LTM. Lastly, because the STM/WM is the "gateway" to the LTM.

8.8.2. *Models*

The main features of STM and WM are described by two functional models (Norris 2017).

8.8.2.1. *Atkinson and Shiffrin's "serial" model*

In this model proposed in 1968 and primarily designed to account for human verbal memory, information which enters a sensory register is first temporarily held in a short-term store (STS) and then *transferred* to a long-term store (LTS). Schematically, these two memories are distinguished 1) by their *storage capacity*, which is very limited in the STS (span) and virtually unlimited in the LTS, and for verbal (or verbalizable) information, 2) by the nature of their coding, which is essentially acoustic/phonological ("tone") in the STS and semantic (meaning) in the LTS and, indirectly, 3) by serial position effects. Thus, if a memory evaluation is carried out immediately after the presentation of a list of n items (e.g. a list of 10 words in humans), it is found that the items placed at the beginning of the list (*primacy effect*) and at the end of the list (*recency effect*) are retained better than the items placed in the middle of the list. The recency effect would correspond to the expression of the STS, and the primacy effect to the expression of the LTS. Serial position effects exist in other species, including monkeys, rats and pigeons.

8.8.2.2. *Baddeley and Hitch's working memory model*

This model, introduced in 1974, combines *components for storing, maintaining and processing information*. It distinguishes between two storage systems. On the one hand, the visuo-spatial sketchpad, which stores visual (objects, forms, patterns) and spatial (position) images for manipulation and is based on mental imagery, and on the other hand, the phonological loop, which operates in the same way for the verbal or verbalizable information it maintains – "recycles" – through a system of subvocal articulatory rehearsing. In this model, the crucial element is the central executive, whose function is to supervise (control and coordinate) the functioning of the other two systems and to link them with the knowledge stored in long-term memory, so as to organize (according to rules), update information and use the result of these manipulations to guide behavior. These functions of the central administrator correspond more broadly to the executive functions involved in other cognitive skills that are controlled by the frontal lobes (see below). Their efficiency would reflect an individual's general intelligence.

8.8.2.3. *Executive functions and cognitive control*

Strategic attention control

This is the ability to focus on the elements that are relevant to the resolution of a problem and ignore those that are not. The most classic test to study this phenomenon is the Stroop test, in which the subject must select the relevant stimulus and inhibit the response automatically generated by the other stimulus. For example, responding "green" (color of ink) to the presentation of the written word red.

Cognitive flexibility

This is the ability to move quickly from one response or strategy to another based on the result produced or the signals received from the environment. The most commonly used test is the Wisconsin card sorting test, in which the subject must match cards according to the color, form or number of items on the card and *change* the type of match chosen as soon as they are warned that it is no longer valid.

Planning

This is the ability to identify and organize the steps and elements necessary to achieve a goal. The most commonly used test is the Hanoi Tower test or other simpler versions, such as the Tower of London (see section 8.6.2.2). These tests are challenging for the WM central executive as the subject must, at all times, remember the previous steps (retrospective memory) and choose the next step from all the remaining steps (prospective memory).

Updating and organizing the content of the STM

As we have seen, these processes operate on the information temporarily held in STM to adapt it to the task at hand. They represent the major feature of the WM. One of the most commonly used tests is the n-back test. The subject is presented with a series of items, usually letters, at the rate of one item every two or three seconds. The subject is instructed to say whether the item currently presented has ever been presented in the "*n-back position*". For example, if n = 3, the subject is instructed to *only* respond if the item presented has been seen three rows before (e.g. R, in the sequence R, J, L, R). The "updating" of each presentation *optimizes the memory load* (in the example, only keeping in mind the last four letters presented).

8.8.3. *Short-term memory in animals*

Short-term memory, which is just the ability to keep information in memory, as is, is certainly a property of all nervous systems capable of learning. However, WM remains rudimentary in animals other than primates and some birds.

8.8.3.1. *Principle of STM testing*

A simple STM test consists of briefly presenting an item and then, after a delay, performing a test in which the animal chooses between that item and a new item. These so-called "delayed recognition" tests use *visual* (*objects*, *colors*, *images*), *auditory*, *olfactory or tactile materials*. Some tests involve the memory of *positions* (so-called "delayed response" tests).

8.8.3.2. *Examples*

In bees, visual information (e.g. vertical blue stripes on a white panel) is retained for a maximum of 8 seconds. In pigeons, 75% of correct responses for the recognition of a color can still be observed after a 10-second delay, whereas in monkeys, the recognition of a 3D object can still be retained 90% of the time after a delay of 30 seconds.

8.8.4. *Cerebral substrates*

8.8.4.1. *Delay cells in monkeys*

By recording neuronal activity in the lateral prefrontal cortex (CPFL) during a delayed response test in monkeys, Fuster and Alexander (1971) discovered many neurons whose characteristic was to maintain high activity (frequency of discharge) during the delay and to return to normal activity as soon as the response was given, suggesting their role in "holding in mind" information needed to make the later response. Subsequent work, notably that conducted by Patricia Goldman-Rakik and her team, confirmed this phenomenon and showed that the location of the delay cells in the monkey's CPFL depended on the nature of the information to be retained: in the *upper* (*dorsal*) part for *positions* (*where*) and *lower* (*ventral*) for the nature of the objects (*what*) (Wilson *et al.* 1993).

8.8.4.2. *Front lobes, executive functions and working memory*

Patients with frontal lobe damage have deficits in executive function and WM. In both humans and monkeys, there is *functional segregation* between the dorsolateral (CPFDL) and ventrolateral (CPFVL) regions of the prefrontal cortex, the former being involved in STM *of positions* and the latter in STM of *objects* (e.g. faces). In addition, while the simple maintenance of a series of letters relies on the activity of the CPFVL, their manipulation (e.g. alphabetizing them) requires the CPFDL, which is also the case in the n-back test (see section 8.8.2.3).

8.9. Conclusion: organization and reconfiguration of the different forms of memory

In *L'homme, ce singe en mosaïque*, Georges Chapouthier (2001, p. 124) writes: "In the evolution of animal species, increasingly complex levels of memory organization are appearing [that] do not replace the most primitive ones, but complement them. It follows that human memory is a sort of patchwork." What are these "complements", why do they appear and what can their role be? There are two possibilities.

First, "evolutionary necessity" may involve adding a system with entirely new functional properties. This is, in the context of evolution, the thesis of functional incompatibility defended by Sherry and Schacter (1987), for whom a system that is perfectly effective in solving one type of problem may be totally unsuitable for solving some others. For example, the memory system specializing in *incremental habit* formation that preserves invariances across episodes cannot, by definition, preserve the *contextual details* that *distinguish* different episodes, and thus the *variance* across those episodes, which is precisely what *episodic memory* allows. Another question is whether episodic memory has a long evolutionary history (Allen and Fortin 2013) or, as Endel Tulving (2002) argues, whether it has recently only emerged in *humans* outside (neo-)Darwinian evolution (Tulving 2002).

They can then allow a finer adaptation or a better control of a pre-existing system. This is for example the case in the habituation to composite stimuli in rats (see section 8.3.1.1) or in the ability to associate the two stimuli (CS and US) of a classical conditioning (CC) despite their temporal separation (so-called "trace" CC) (see section 8.4.1). In both these situations, the mammalian hippocampus is required. But, concerning CC, one cannot fail to note that evolution has short-circuited it as soon as it was a question of conditioning a taste aversion that not only tolerates but "requires" that a certain amount of time elapses between the perception of the taste of the food consumed and the visceral discomfort that follows. Add to this the fact that CC exists in sea anemones that do not have a "centralized" nervous system, and it is clear that the singularity of CC "masks" an astonishing diversity of expression and "control". But what does the hippocampal system "do" in this conditioning? For Clark and Squire (1998), the trace CC of the palpebral reflex is only established if and because the subjects (in the experiment, humans) "became aware" of the relationship between the CS and the US. However, many examples show that the hippocampus is not systematically associated with that which is "conscious" or "explicit", but that its role is rather to construct *relational representations* (see section 8.7.1.2) whose acquisition and expression may be unconscious or implicit (see, for example, Henke 2010). This is also why, when the information to be retained is relational in nature, the hippocampus would contribute to the functioning of *short-term memory* (STM) (see, for example, Olson *et al.* 2006), which brings up the problem of independence between STM and LTM (see, for example, Ranganath and Blumenfeld 2005), and thus the problem of STM preservation in amnesic syndrome (Yee *et al.* 2014). Thus, it can be observed that the hippocampus is far from remaining "assigned" to the place shown in the memory organization depicted in Figure 8.1.

In the same vein, another interesting example is the amygdala (Jaffard 2016), which is necessary for the establishment of *conditioned fear*. Indeed, by its position, this subcortical structure associates the information transmitted by a simple somatic-sensory stimulus or a particular context (CSs) with which it conveys the perception of the aversive event (US). After conditioning, the amygdala then triggers the *physiological* (e.g. adrenaline release) and *behavioral* (e.g. immobilization) fear responses triggered by the CSs. It can be noted in passing that context-conditioned fear cannot be established without the "collaboration" of the hippocampus, the only one capable of constructing a representation of it. It must also be noted that, without the hippocampus, fear conditioning to a simple CS in humans is expressed without any episodic memory of the initial situation (Bechara *et al.* 1995). That being said, Ledoux and Pine (2016) consider that the subjective (explicit) sensation of fear that accompanies the (implicit) physiological and behavioral responses to fear in humans involves the addition of frontal-parietal cortical "cognitive circuits" related to attention and working memory. However, there is (much) more. Before the research on fear conditioning began, it was known that the amygdala also plays a key role in *facilitating memory* through the emotional arousal, both positive or negative, that accompanies or follows an experience. Essentially, it has been established that this emotional awakening leads to *activation of the amygdala* which, through its projections, facilitates *memory consolidation* by modulating the neuroplasticity of the *hippocampus*, *dorsal striatum* and *neocortex*. Thus, in addition to its role in the "storage" of (implicit) fear conditioning, the amygdala facilitates the storage of different forms of memory in other brain structures (McGaugh 2018).

We started from a "vertical" (or parallel) organization of the different forms of memory (see Figure 8.1), which allowed us to give an "ordered" description of them by moving from the simplest (habituation on the right) to the most complex (episodic memory on the left). The architecture of this organization can certainly be preserved, but, as we have just seen, we must admit that the "brain modules" are not functionally fixed, but also operate "horizontally". Although the representations elaborated by each of the "modules" and their "behavioral output" are often doubly dissociable (e.g. hippocampus-memory versus neostiatum-habit, see section 8.6.1), recent data show that they can reconfigure or associate temporarily according to the constraints imposed by the situation. For example, if the same animal learns two types of strategies ("response" and "spatial") that are used on "instruction" (if ... then), the dorsal striatum *and* the hippocampus are *no longer* functionally dissociable (Ferbinteanu 2018).

8.10. References

Allen, T.A., Fortin, N.J. (2013). The evolution of episodic memory. *Proc. Natl. Acad. Sci. USA*, 110, 10379–10386.

Balleine, B.W., O'Doherty, J.P. (2010). Human and rodent homologies in action control : Corticostriatal determinants of goal-directed and habitual actions. *Neuropsychopharmacology Reviews*, 35(1), 48–69.

Bayley, P.J., Squire, L.R. (2005). Robust habit learning in the absence of awareness and independent of the medial temporal lobe. *Nature*, 436(7050), 550–553.

Bechara, A., Tranel, D., Damasio, H., Adolphs, R., Rockland, C., Damasio, A.R. (1995). Double dissociation of conditioning and declarative knowledge relative to the amygdala and hippocampus in humans. *Science*, 269(5227), 1115–1118.

Bergson, H. (1896). Matière et mémoire [Online]. Available at: http://classiques.uq ac.ca/classiques/.

Bohbot, V.D., Lerch, J., Thorndycraft, B., Iaria, G., Zijdenbos, A.P. (2007). Gray matter differences correlate with spontaneous strategies in a human virtual navigation task. *J. Neurosci.*, 27(38), 10078–10083.

Chapouthier, G. (2001). *L'homme, ce singe en mosaïque*. Odile Jacob, Paris.

Clark, R.B. (1960). Habituation of the polychaete Nereis to sudden stimuli. I. General properties of the habituation process. *Anim. Behav.*, 8(1–2), 82–91.

Clark, R.E., Squire, L.R. (1998). Classical conditioning and brain systems: the role of awareness. *Science*, 280(5360), 77–81.

Clayton, N.S., Dickinson, A. (1998). Episodic-like memory during cache recovery by scrub jays. *Nature*, 395(6699), 272–274.

Colwill, R.M., Rescorla, R.A. (1985). Postconditioning devaluation of a reinforcer affects instrumental responding. *J. Exp. Psychol. Anim. Behav. Process.*, 11(1), 120–132.

Eichenbaum, H. (2013). Memory on time. *Trends Cogn. Sci.*, 17(2), 81–88.

Ferbinteanu, J. (2018). Memory systems 2018 – Towards a new paradigm. *Neurobiol. Learn. Mem.*, 157, 61–78.

Fleischman, D.A., Wilson, R.S., Gabrieli, J.D.E., Schneider, J.A., Bienias, J.L., Bennett, D.A. (2005). Implicit memory and Alzheimer's disease neuropathology. *Brain*, 128, 2006–2015.

Fuster, J.M., Alexander, J.E. (1971). Neuron activity related to short-term memory. *Science*, 173(3997), 652–654.

Garcia, J., Kimeldorf, D.J., Koelling, R.A. (1955). Conditioned aversion to saccharin resulting from exposure to gamma radiation. *Science*, 122(3160), 157–158.

Gilbert, C.D., Sigman, M., Crist, R.E. (2001). The neural basis of perceptual learning. *Neuron*, 31(5), 681–697.

Hamilton, T.J., Myggland, A., Duperreault, E., May, Z., Gallup, J., Powell, R.A., Schalomon, M., Digweed, S.M. (2016). Episodic-like memory in zebrafish. *Anim. Cogn.*, 19(6), 1071–1079.

Hartley, T., Maguire, E.A., Spiers, H.J., Burgess, N. (2003). The well-worn route and the path less traveled: Distinct neural bases of route following and wayfinding in humans. *Neuron*, 37(5), 877–888.

Henke, K. (2010). A model for memory systems based on processing modes rather than consciousness. *Nat. Rev. Neurosci.*, 11(7), 523–532.

Honey, R.C., Watt, A., Good, M. (1998). Hippocampal lesions disrupt an associative mismatch process. *J. Neurosci.*, 18(6), 2226–2230.

Jaffard, R. (2011). La mémoire déclarative et le modèle de Squire. *Rev. Neuropsychol.*, 3(2), 83–93.

Jaffard, R. (2016). Le rôle clé de l'amygdale. In *Mémoire et émotions*, Eustache, F. (ed.). Le Pommier, Paris.

James, W. (1890). The principles of Psychology [Online]. Available at: http://psychclassics.yorku.ca/James/Principles/index.htm.

Karni, A., Sagi, D. (1993). The time course of learning a visual skill. *Nature*, 365(6443), 250–252.

LeDoux, J.E., Pine, D.S. (2016). Using neuroscience to help understand fear and anxiety : a two-system framework. *Am. J. Psychiatry*, 173(11), 1083–1093.

Lerch, J.P., Yiu, A.P., Martinez-Canabal, A., Pekar, T., Bohbot, V.D., Frankland, P.W., Henkelman, R.M., Josselyn, S.A., Sled, J.G. (2011). Maze training in mice induces MRI-detectable brain shape changes specific to the type of learning. *NeuroImage*, 54(3), 2086–2095.

Loftus, E.F. (2003). Make-believe memories. *Am. Psychol.*, 58(11), 867–873.

Maine de Biran, F.-P. (1799). Influence de l'habitude sur la faculté de penser [Online]. Available at: http://classiques.uqac.ca/classiques/.

McGaugh, J.L. (2018). Emotional arousal regulation of memory consolidation. *Curr. Opin. Behav. Sci.*, 19, 55–60.

Milner, B. (1962). Les troubles de la mémoire accompagnant des lésions hippocampiques bilatérales. In *Physiologie de l'hippocampe*, Colloques Internationaux no. 107. CNRS, Paris.

Murray, E.A., Wise, S.P., Graham, K.S. (2018). *The Evolution of Memory Systems*. Oxford University Press, Oxford.

Norris, D. (2017). Short-term memory and long-term memory are still different. *Psychol. Bull.*, 143(9), 992–1009.

Olson, I.R., Page, K., Moore, K.S., Chatterjee, A., Verfaellie, M. (2006). Working memory for conjunctions relies on the medial temporal lobe. *J. Neurosci.*, 26(17), 4596–4601.

Pavlov, I.P. (1928). *Lectures on Conditioned Reflexes*. International Publishers, New York.

Poldrack, R.A, Clark, J., Paré-Blagoev, E.J., Shohamy, D., Creso Moyano, J., Myers, C.M., Gluck, M.A. (2001). Interactive memory systems in the human brain. *Nature*, 414(6863), 546–550.

Ranganath, C., Blumemfeld, R.S. (2005). Doubts about double dissociations between short- and long-term memory. *Trends Cogn. Sci.*, 9(8), 374–380.

Rescorla, R.A. (1988a). Behavioral studies of pavlovian conditionning. *Annu. Rev. Neurosci.*, 11, 329–352.

Rescorla, R.A. (1988b). Pavlovian conditioning. It's not what you think it is. *Am. Psychol.*, 43(3), 151–160.

Rescorla, R.A., Wagner, A.R. (1972). A theory of Pavlovian conditioning: variations in the effectiveness of reinforcement and non reinforcement. In *Classical Conditioning II: Current Research and Theory*, Black, A.H., Prokasy, W.F. (eds.). Appleton-Century-Crofts, New York, 64–99.

Schacter, D.L., Benoit, R.G., Szpunar, K.K. (2017). Episodic future thinking: mechanisms and functions. *Curr. Opin. Behav. Sci.*, 17, 41–50.

Schrödinger, E. (1990). *L'esprit et la matière*. Le Seuil, Paris.

Schultz, W. (2016). Dopamine reward prediction error coding. *Dialogues Clin. Neurosci.*, 18(1), 23–32.

Semon, R. (1921). *The Mneme*. The MacMillan Company, New York.

Sherry, D.F., Schacter, D.L. (1987). The evolution of multiple memory systems. *Psychol. Rev.*, 94(4), 439–454.

Siegel, S. (2016). The heroin overdose mystery. *Curr. Dir. Psychol. Sci.*, 25(6), 375–379.

Squire, L.R., Dede, A.J. (2015). Conscious and unconscious memory systems. *Cold Spring Harb. Perspect. Biol.*, 7.

Squire, L.R., Kandel, E.R. (2001). *La mémoire. De l'esprit aux molécules*. De Boeck Université, Brussels.

Straube, B. (2012). An overview of the neurocognitive processes involved in the encoding, consolidation and retrieval of true and false memories. *Behav. Brain Funct.*, 8, 35.

Thorndike, E.L. (1911). Animal Intelligence [Online]. Available at: http://psychclassics.yorku.ca/Thorndike/Animal/.

Tolman, E.C. (1932). *Purposive Behavior in Animals and Men*. The Century Co, New York.

Tolman, E.C. (1948). Cognitive maps in rats and men. *Psychol. Rev.*, 55(4), 189–210.

Tulving, E. (1972). Episodic and semantic memory. In *Organization of Memory*, Tulving, E., Donaldson, W. (eds). Academic Press, New York.

Tulving, E. (2002). Episodic memory: From mind to brain. *Annu. Rev. Psychol.*, 53, 1–25.

Vargha-Khadem, F., Gadian, D.G., Watkins, K.E., Connelly, A.,Van Paesschen, W., Mishkin, M. (1997). Differential effects of early hippocampal pathology on episodic and semantic Memory. *Science*, 277(5324), 376–380.

Warrington, E.K., Weiskrantz, L. (1968). New method of testing long-term retention with special reference to amnesic patients. *Nature*, 217(5132), 972–974.

White, N.M., Packard, M.G., McDonald, R.J. (2013). Dissociation of memory systems : The story unfolds. *Behav. Neurosci.*, 127(6), 813–834.

Wilson, F.A., Scalaidhe, S.P., Goldman-Rakic, P.S. (1993). Dissociation of object and spatial processing domains in primate prefrontal cortex. *Science*, 260(5116), 1955–1958.

Yee, L.T., Hannula, D.E., Tranel, D., Cohen, N.J. (2014). Short-term retention of relational memory in amnesia revisited: accurate performance depends on hippocampal integrity. *Front. Hum. Neurosci.*, 8, 16.

Zeithamova, D., Schlichting, M.L., Preston, A.R. (2012). The hippocampus and inferential reasoning: building memories to navigate future decisions. *Front. Hum. Neurosci.*, 6, 70.

9

The Construction of Sensory Universes

Dalila BOVET
University of Paris Nanterre, France

9.1. Introduction

In this chapter, we are interested in how animals (including humans) perceive the world and construct their sensory universes by combining the information provided to them by their different senses. When studying animals, it is indeed important to understand what they perceive in their environment and how this affects their interactions with that environment. Sensory receptors convert the energy of a *stimulus* into nerve impulses for transmission to the central nervous system. In humans, we classically speak of the five senses; but in biology, there are more sensory receptors, which can be classified according to the nature of the *stimuli* they convert: chemical (or chemoreception: taste and smell), mechanical (mechanoreception: touch, hearing and balance), electromagnetic (which includes photoreception, therefore vision, but also electro-reception: sensitivity to electric current, and magnetoreception: sensitivity to the earth's magnetic field), and thermal (thermoreception: sensitivity to heat). Animals also possess internal senses that use the same modalities. However, we will focus here on the external senses that allow us to perceive external *stimuli*.

The relative importance of each sensory modality and the organization of the information perceived by the different senses varies from one species to another – a variation that can inform us about the ecology and evolution of different species.

The Explosion of Life Forms,
coordinated by Georges CHAPOUTHIER and Marie-Christine MAUREL. © ISTE Ltd 2020.

We will see that many animals can detect *stimuli* that are not perceptible by human sensory systems. Certain types of receptors don't exist in humans. For others, it is a difference in degree of sensitivity; generally, the prefixes "ultra" and "intra" are used, the former for *stimuli* above the range that humans can detect and the latter for *stimuli* that are below this range. This chapter will not be an exhaustive review of animal sensory systems, but rather an overview to give an idea of their immense diversity.

9.2. Chemoreception

Chemoreception is the process by which organisms detect chemical *stimuli*. Chemoreception is probably the most primitive sensory system, observed in almost all animal species, as well as in micro-organisms such as bacteria. A distinction is generally made, especially for land animals, between the sense of taste (for chemicals in solution in water, informing on the flavor of potential food) and the sense of smell (for volatile airborne chemicals, informing on distant *stimuli*).

9.2.1. *Taste*

In vertebrates, taste is detected by the taste buds. Most of these taste buds are found on the surface of the tongue, associated with protruding structures. Each of these taste buds contains between 50 and 150 taste receptor cells, with which incoming chemicals interact to produce electrical signals to the brain. Taste buds are found on the tongue and other areas of the mouth in land vertebrates, but also on the lips, flanks and tails of some fish, including on catfish barbels. Taste receptor cells in vertebrates are continually renewed throughout the life of the organism.

Vertebrates can sense at least five different taste modalities: sweet, salty, sour, bitter and umami. These modalities provide animals with information about the nutrient content and potential hazards of the food they eat: sweet foods are naturally attractive because sugars serve as the main source of energy for animals. The bitter taste prevents the ingestion of toxic substances such as alkaloids, synthesized by various plants (specifically to protect against ingestion by animals). Animals do not sense the taste of proteins as such but can taste their constituents, the amino acids. Indeed, some amino acids have a sweet taste, others have a sour taste, and the taste of umami (from the Japanese word meaning "tasty"), evoked notably by meat, mushrooms, or certain fermented products, is a response to glutamic acid and its derivatives. The salty and acidic tastes are additional protective mechanisms to ensure the internal balance of sodium or acid-base respectively. Animals must indeed eat sufficient salt, while avoiding excesses that would have harmful effects on cellular metabolism. Natural foods contain many different chemicals; for example, the taste of a grapefruit can stimulate all types of receptors to different degrees.

Taste receptor cells in vertebrates appear to be generally sensitive to all of these modalities because the basic food selection needs of most animals are similar. However, sensitivity to some modalities may be lower or even absent in some species: felines, being strict carnivores, do not sense sweetness. Herbivores are better at distinguishing different bitter compounds than carnivores, but they also tolerate them better, presumably because bitter substances are common in their diet and rejecting them all would make it difficult to get enough to eat. In addition, herbivores often have defense mechanisms against toxins, such as fermentation by rumen bacteria in ruminants. In dolphins, who swallow fish without tasting it, most taste genes are no longer functional.

On average, the human tongue has 8,000 taste buds (but individual differences are very large). Cows, who are herbivores, break all records with about 25,000 taste buds, while pigs, who are omnivores, have an average of 15,000 taste buds. Carnivores have the fewest taste buds (around 800 in cats). Birds have fewer taste buds than mammals (about 30 in hens, but several hundred in parrots).

Like vertebrates, arthropods seem to have very distinct senses of taste and smell. Their taste cilia are primarily found on the food appendages associated with the mouth, but located outside the mouth. Taste receptors are also present in the antennae of ants, for example. Many arthropods, such as butterflies and some species of flies, have taste receptors on their legs, allowing the animal to taste what it is walking on. In some species, similar receptors are scattered on the body surface and may also be present on the egg-laying apparatus, making it easier to recognize host plants on which to lay eggs. Drosophila also have taste senses on their wings, which help them better orient themselves to food sources while flying.

Insect taste receptors also respond primarily to food-related chemicals. The sensitivity of the cells varies according to the nature of the insect's food. In most species that feed on plants, the receptors are sensitive to sugars, amino acids, inorganic salts and various compounds produced by plants that generally inhibit feeding. These four categories therefore roughly correspond to the four sweet, sour, salty and bitter modalities. Blood-sucking insects have receptor cells that are sensitive to adenine nucleotides (ADP and ATP) carried by blood and mosquitoes or blowflies have cells that respond to very low salt concentrations. Some phytophagous insects have taste receptor cells that are specialized in the perception of host-specific chemicals. For example, plants of the cabbage family (cruciferous vegetables) are characterized by a class of compounds called glucosinolates, and some insects that feed on cruciferous plants have receptors that respond only to glucosinolates. The sense of taste can also be used in some insects, such as ants, to detect pheromones on the cuticle of other insects.

9.2.2. *Smell*

The olfactory system therefore allows us to perceive chemicals coming from outside. These signals do not fall into categories as clear as the taste receptor system, and most organisms have the ability to distinguish between hundreds or even thousands of smells, even for very similar chemical structures.

In land vertebrates, receptor olfactory cells are neurons, forming an olfactory epithelium in the nasal cavity. The receptor ends of these cells have cilia that extend into the mucus covering the wall of the nasal cavity. When an odorous molecule reaches a receptor, it binds to a specific protein on the plasma membrane. The olfactory receptor then produces an action potential that directly reaches the olfactory bulb of the brain, where the olfactory information is processed. With up to 1,000 different types of receptor proteins, the number of possible combinations is enormous. This wide range of combinations gives animals their extraordinary ability to distinguish thousands of smells.

The relative size of the olfactory epithelium reflects the importance of olfaction in the lives of different animals. In some dogs, the olfactory epithelium has an area of about 170 cm^2, or one billion olfactory receptor cells in total; in cows, the olfactory epithelial surface area is only 1 to 4 cm^2 and the number of cells is less than 30 million. The human olfactory epithelium covers 5 to 10 cm^2 and has 10 to 40 million olfactory receptor cells. The family of genes that regulate the production of olfactory receptors is common to all vertebrates. However, the number of active olfactory receptor genes is variable. In mice, which have a highly developed sense of smell, most of the near 1,000 olfactory genes are expressed (in other words, they produce receptor proteins). However, in primates, up to 70% of the olfactory receptor genes, although still identifiable, are non-functional pseudogenes. All olfactory genes in dolphins are non-functional.

The olfactory system of fish is independent of the respiratory and oral systems. The concept of the sense of smell is moreover quite widespread in fish and amphibians. This is because in addition to olfactory receptors located in the nasal cavity, chemoreceptor cells are dispersed on the skin of the fish. As in land vertebrates, the number of olfactory receptor cells can be very large: up to 10 million.

Many invertebrates have olfactory chemoreceptor cells contained in structures called "sensilla". These sensilla may be present all over the body, including on the legs, the cerci (appendages at the end of the abdomen) and the wing margins. The number of chemoreceptor cells varies greatly from one species to another. *Caenorhabditis elegans*, a small worm living in the soil, only has 34. In insects, the length or complexity of the antennae reflects the number of olfactory senses, hence the feather-like antennae of most male moths that enable them to perceive the pheromones of females. One kind of moth, the male tobacco moth, can have 100,000 to 300,000 receptor cells on one antenna. Some crabs have a similar number of olfactory receptor cells on their antennae.

Arthropods are generally able to react to, and differentiate between, a wide range of chemical compounds, including pheromones and food-related odors. The characteristic odors of many plants can be perceived and, depending on the species, attract or repel insects. Many species of phytophagous insects can perceive metabolic by-products produced by green plants. Blood-sucking insects perceive carbon dioxide, which can be an important indicator of the presence of a host. Laboratory experiments have shown that arthropods, like vertebrates, also perceive a wide range of odors that do not seem relevant to their lifestyle. This ability is probably necessary to develop associations between odors and important but unpredictable factors in the lives of animals.

Single-celled organisms are also capable of chemoreception, which allows them to remain in favorable chemical environments and to move away from environments that are not. For example, *Escherichia coli* bacteria accumulate in areas with high concentrations of sugars and certain amino acids. *Parameciums* accumulate in areas with high concentrations of folate or biotin, compounds released by the bacteria on which these ciliated protozoa feed. On the other hand, *parameciums* disperse in the presence of quinine or potassium hydroxide. The number of different types of receptor proteins is limited in unicellular organisms versus multicellular organisms, probably because the space available on the surface of a single cell is reduced.

9.3. Mechanoreception

Sensitivity to mechanical *stimuli* is present in most animals. There are many varieties of mechanical receptors in animals, but the main ones are the receptors that mediate touch and the hair cell receptors that mediate hearing.

9.3.1. *Touch*

In vertebrates, tactile receptors are present in the skin. Local pressure on the body surface causes deformation of parts of the receptor and the appearance of a receptor potential. This receptor potential generates action potentials towards the brain when it reaches sufficient strength. In insects, it is tactile cilia called "sensilla" that affect the receptor potential. Some insects have such sensitive sensilla that they can detect slight air movements and even variations in atmospheric pressure. For example, potato aphids reduce their reproductive behavior when it senses rain.

Many vertebrates and invertebrates can accurately locate points of tactile stimulation on the body surface. Humans, for example, can distinguish between two similar pointed *stimuli* when the points are separated by as little as 1 millimeter at the tip of the tongue, but this sensitivity is much less on other parts of the body. Many animals can also distinguish tactile *stimuli* that differ qualitatively, such as a rough surface and a smooth surface. This ability can even be observed in some ciliated protozoa. The long antennae of many arthropods (e.g. crayfish) and the vibrissae (commonly called "whiskers") of the muzzles of nocturnal mammals (e.g. cats or rats) are used for tactile detection of objects close to the animal's body. In laboratories, rats can use their vibrissae alone to assess the width of a corridor (to an extent of 3 millimeters), the shape of a biscuit, or to differentiate between smooth and rough textures with similar accuracy to a human fingertip. Shrews also use their vibrissae to recognize prey in the dark with its shape. It is also vibrissae that allow seals to follow traces of turbulence in the water left by fish. The star-nosed mole, which lives underground in the dark, has tentacles on its snout that allow it to identify and capture small prey in 120 milliseconds.

9.3.2. *Lateral lines*

Fish and amphibians have special mechanoreceptors in sensory organs called "lateral lines". These are a series of channels in the skin that are open to the surrounding water and allow them to perceive water currents caused by the animal itself, or by other animals. These canals are equipped at regular intervals with clusters of hair cells, each with a cup. The cup is moved by the movement of the water, creating action potentials. These organs are sensitive to minimal and local water movement, especially that produced by other animals moving in the water. Thus, the lateral lines allow a kind of remote touch, to perceive and locate moving *stimuli* (including prey, predators, or fellow fish) before physical contact occurs. These organs are particularly useful for swimming in schools. However, they can only perceive close *stimuli*. Lateral lines also allow fish to orient themselves in relation to water currents, and even to perceive low frequency waves (low sounds).

In the embryo, the mechanoreceptors of the lateral lines develop from the same structure as the sensory cells of an inner ear structure, called the "labyrinth", which is similar in all vertebrates, but is used more for balance than for hearing in fish. In amphibians such as frogs, the lateral lines disappear during the metamorphosis of tadpoles. Land vertebrates do not have lateral lines; only the deeply located labyrinthine sensory organs remain to allow hearing.

9.3.3. *Hearing*

The hearing sense is less common than the other senses in the animal kingdom, since it is mainly limited to vertebrates and arthropods. All vertebrates have at least one inner ear inherited from their common fish ancestors and all have some hearing ability.

In mammals, sound waves enter the ear and travel through the ear canal. When the waves reach the tympanic membrane, they vibrate the membrane and the chain of auditory ossicles attached to it. This stimulates the hair cells of the organ of Corti to send nerve impulses to the brain. The human ear is sensitive to sounds with frequencies ranging from 20 hertz to 20 kilohertz.

Ultrasounds (in other words, sounds at a frequency of 20 kilohertz or more) are produced or detected by various species. Many rodents communicate by ultrasound. Bats and cetaceans detect sounds at very high frequencies that can exceed 100 kilohertz. These animals use echolocation, both for navigation and for capture of prey: they emit a high-frequency call and listen to the echo. The time it takes for the echo to return tells them the distance to the target, the frequency offset gives the relative speed of the transmitter and receiver, and the frequency spectrum of the returned echo contains information about the size and texture of the target. Dolphins and bats are very distant species on the evolutionary scale, but their echolocation is the product of convergent evolution: the same genes have mutated independently to promote echolocation (in darkness for bats, in murky waters for cetaceans). This is notably the case of a coding gene called the prestine, a protein essential to hearing and sound amplification. This protein acts as an amplifier in the inner ear by modifying the length of the sensory cilia in the hair cells. It therefore plays a major role in the molecular mechanism of echolocation. Bats generally emit few vocalizations during navigation, but when they detect an insect, the rate of emission accelerates to give a "capture buzz" as the bat approaches its prey. The frequency range of bat echolocation calls varies from 10,000 to 200,000 hertz, depending on the species. Many insects have developed countermeasures against echolocation, including the ability to perceive high frequencies, combined with a powerful diving strategy on the ground when sounds emitted by bats are detected. Some moths, such as peppered moths, produce high-frequency vocalizations themselves to create acoustic confusion. Since high-frequency calls only cover short distances, a bat or dolphin's perception of the world is likely affected by the distances over which these sounds travel. This contrasts with the visual perception of the environment, which is made over much longer distances.

Other species, however, prefer low frequencies (thus low sounds). Elephants produce sounds lower than 20 hertz (called "infrasound") to communicate with the other members of their herd. These low-frequency sounds travel great distances in the air and across the ground, allowing groups of elephants to stay in contact over several dozen kilometers. Elephants also sense ground vibrations through pads under their legs.

The hearing sensitivity of reptiles is low and reduced at low frequencies, and even infrasound. As a result, snakes are virtually deaf or only sense very low frequency sounds. They mainly pick up ground vibrations through the bones of their skull and, to a lesser degree, the rest of their skeleton. Snakes therefore do not hear the snake charmer's flute, but sense ground vibrations (because the snake charmer taps his foot), and they also respond to visual *stimuli* represented by the movements of the flute and the snake charmer. Similarly, turtles only hear the lowest frequencies (10 to 182 hertz in Hermann turtles, for example). The turtle hearing system is supplemented by the conduction of the shell bones to detect vibrations in the immediate environment, which can be particularly useful when the turtle has its head in the shell, making sound reception even worse. Despite these limitations, the turtle's hearing system makes it possible to detect the approach of potential predators and to communicate with their congeners: the vocalizations of male Hermann turtles allow them to keep other males away, but are also reliable indicators of their condition, used by the females to select them.

In arthropods, the ability to detect sounds seems to have evolved independently on several occasions, resulting in a wide variety of hearing structures and their locations, which can be found on almost any part of the body: for example, eardrums open on the tibia, near the knee, in grasshoppers and crickets. In locusts, they are located on the first segment of the abdomen. The auditory sensitivities of arthropods are also highly variable, ranging from 10 hertz to more than 100 kilohertz. Insects produce sounds in different contexts, often the same as in vertebrates, including intraspecific aggression and the search for mating partners. In cicadas, for example, the sound is only produced by males, in the abdomen. The abdomen contains two semi-rigid cymbals which, activated by powerful muscles, produce a powerful snap at a frantic rate, several hundred times per second. The abdomen is almost hollow, serving as a sound box; the organs are compressed at the extremities. A species of

Australian cicada, *Cyclochila australasiae*, has been recorded at 120 decibels, the equivalent of a plane taking off. No man-made device can produce such a loud sound using so little space and energy. The sound performance of the water bug (*Micronecta micronecta scholtzi*) is also surprising: this aquatic insect has been nicknamed the "singing penis" because it manages to produce a sound of 99 decibels by rubbing its penis against its abdomen.

As noted above, the sense of hearing, as the perception of sound vibrations through a specialized organ, does not exist in non-arthropod invertebrates. In general, these species do not produce sounds and therefore do not communicate through this channel. However, they are frequently able to detect vibrations produced by prey or predators through different substrates, such as plants, cobwebs and soil.

9.4. Electromagnetoreception

9.4.1. *Vision*

Many animals use vision to determine the layout of their environment. The organs used for this are very variable according to taxa. Some are simple clusters of cells that are sensitive to the intensity and direction of light, while other species have complex organs that enable them to represent the objects around them. These include compound eyes (found in insects, crustaceans and some mollusks), and the so-called "camera eyes" (by analogy with cameras) of vertebrates and cephalopods, which project inverted images onto a retina. Although many other characteristics of vision (resolution, distance, etc.) vary from taxon to taxon, we will limit ourselves here to an overview of color vision.

Vision is based on cells called "photoreceptors", which contain pigment molecules capable of absorbing light waves. Similar genes for these photoreceptors are found in animals belonging to very distant taxa, such as platyhelminths (or flat greens), annelids (or roundworms), arthropods and vertebrates. This suggests a common genetic origin of photoreceptors that appeared in the first bilaterally symmetric animals.

The visual pigments of vertebrates consist of a molecule capable of absorbing light (the retinal), bound to a membrane protein called "opsin". When the retinal absorbs a photon of light, it changes its configuration, triggering a series of molecular reactions that lead, in a few milliseconds, to a nerve impulse that is transmitted to the brain. It can be assumed that the opsins of the primitive photoreceptors only reacted to the amount of light. Vision, therefore, most probably first appeared in grey levels, depending on the intensity of the light. Color provides additional information and makes it possible to better define the contours of objects (a red fruit on green foliage, for example). In the course of evolution, pigments capable of detecting specific ranges of wavelengths would have appeared to enable color vision. Thus, different opsins with different amino acid sequences allow an organism to have receptors that are more responsive to certain wavelengths. For example, in humans, the three classes of photoreceptor cells called "cones", which provide color vision in daylight, are particularly sensitive to blue (419 nanometers), green (531 nanometers) and red (558 nanometers), respectively. Several types of cones can be stimulated at the same time, and their combination gives us all the colors we perceive.

The first fish had four different opsins to detect four spectra: red, green, blue and ultraviolet light. They were therefore tetrachromates (as are still many fish today), and this was probably the case for dinosaurs, the ancestors of today's birds, which are also mostly tetrachromates. On the other hand, the first mammals were nocturnal and their visual system has lost this complexity. The majority of today's mammals are dichromates. For example, dogs have only two categories of opsins, one of which mainly reacts to blue tones and the other to green-yellow tones. Dogs therefore see yellow and blue in much the same way as humans do, but they perceive greyish-brown instead of red, dark yellow instead of orange, dark blue instead of purple, etc. Most primates (including humans) have regained trichromatic vision that allows them to discriminate between red and green, presumably to encourage the discovery of ripe fruit and tender leaves, while many rodent species are sensitive to ultraviolet light (which could allow them to detect the urinary markings of their congeners).

As for birds, their elaborate perception of color is able to help them detect their food, but is also important in their social relationships: in many bird species, while females have dull, cryptic plumage to be less visible to predators, males have very colorful plumage, which is linked to reproduction. These bright colors and the presence of symmetrical patterns are generally correlated with an effective immune system and good general condition. These are therefore clues that can be used by females to select the best males. UV vision also gives birds important information: for example, male and female blue tits, which are almost identical for human eyes, are easily differentiated for members of their species, because the reflectance of their plumage in UV light is different, and females prefer males that reflect more UV.

Many insects, such as bees, are trichromates and have photoreceptors that are sensitive to UV, blue and green light. This vision apparently precedes the appearance of flowering plants. Many flowering plant species are pollinated by insects, so it is likely that it is the colors of the flowers that have evolved to adapt to the sense of color in insects. When these flowers are examined under UV light, light and dark patterns are visible. These patterns guide the bees to the part of the flower where they will find nectar, while at the same time promoting the passage of the bees through the structures containing the pollen. Researchers have also found that in areas where bees do not pollinate flowers regularly, the number of UV-reflective flowering plant species is much lower than in other areas where bees do pollinate.

Surprisingly, some species have pentachromatic vision and more: stomatopods (mantis shrimp) and butterflies have up to twelve spectral sensitivities! However, discrimination of almost any color can be achieved with three spectral channels in the wavelengths visible to humans: from 400 to 700 nanometers. To enable perception in UV light, four spectral sensitivities are required. Above this number, the gain of information is increasingly small. It is likely that stomatopods and butterflies, unlike vertebrates, do not combine the information provided by the different channels. Instead, each channel would be triggered separately and stimulate neural mechanisms (innate or learned) that are associated with a given color. Furthermore, butterflies and other insects, including bees, have their eyes subdivided into different zones, with the downward-facing zones being specialized for foraging, while the forward-facing or upward-facing zones are used to detect a mate or prey in the sky. Other dorsal areas of the eyes mostly contain UV receptors and are associated with navigation, due to the polarization of sunlight in the sky.

Some animals can detect infrared radiation (wavelengths above 700 nm). However, animals that detect wavelengths in infrared do so with receptors that detect heat, rather than with photoreceptors. We will therefore discuss this in the section devoted to thermoreception.

9.4.2. *Electroreception*

Two families of freshwater fish, the mormyrids from Africa and the gymnotids (electric eels) from South America, have independently developed a very special sense for detecting objects in their environment and for communication. These fish generally inhabit turbid rivers, such as the Amazon or the Nile, where vision is very limited. They have an organ in their tail, derived from a nerve or muscle, which sends out small electrical discharges into the surrounding water. They also have a set of receptors, derived from lateral lines and located on the front part of their body, which detect the electric field produced by the tail organs. Objects in the vicinity of the fish distort this field and the changes are detected and interpreted in terms of the location and electrical properties of the objects over a distance of a few meters. This sense is also used in aggressive and sexual communication. Other fish such as sharks, rays, or catfish have electroreceptors, but no electrical organs, and they also use electroreception passively to detect electric fields that are the result of neuromuscular activity of prey hidden in the sand or mud at the bottom. Sharks, for example, can perceive variations of 0.01 microvolts per centimeter along the surface of their body.

9.4.3. *Magnetoreception*

Detection of the Earth's magnetic field exists in a variety of animal species, including some vertebrates, mollusks and insects. Many migratory birds, as well as carrier pigeons, can orient themselves even in the most overcast weather or fog using this sense. Sea turtles and salmon use magnetoreception to return to where they were born in order to lay their eggs (in combination with scenting, which allows them, when they are nearby, to recognize their native beach or river more accurately). It is also likely that foxes are able to combine the sounds of rodents moving several meters beneath the snow with the slope of the Earth's magnetic field, and therefore deduce the distance to their prey in order to dive into the snow at the appropriate depth. Some researchers have suggested that dogs use the north-south axis for defecation and that cows line up along magnetic field lines, but these results have yet to be confirmed.

Magnetoreception has not yet been fully clarified, but probably relies on two mechanisms: the first is based on grains of magnetite, an iron oxide found in the skulls of many vertebrates (we have already mentioned salmon, sea turtles and carrier pigeons, but it is also found in the human brain), as well as in the teeth of certain mollusks, the abdomen of bees and in certain unicellular organisms known as "magnetotactics", which orient themselves according to the Earth's magnetic field. These grains of magnetite are believed to act as micro-compasses. The second mechanism would be based on cryptochromes, blue-light photoreceptors particularly present in the eyes of birds. The Earth's magnetic field could affect these cryptochrome reactions in a way that would influence bird visual systems, through the integration of geomagnetic and light information.

9.4.4. *Thermoreception*

Many animal species are able to detect heat in the environment. Some species of snakes (including pythons, boas, vipers and rattlesnakes) have specialized systems called "sensory dimples" to perceive and measure the infrared energy emitted by their prey. The inside of each dimple has a thin membrane. When infrared radiation heats up this membrane, the ion channels open at a certain threshold, generating an electrical signal. In mammals, the same channels are activated by chemical irritants such as mustard or certain acids.

These dimples allow predatory snakes to hide in a secluded area and wait for infrared-emitting prey to pass nearby. These receptors can also be used by snakes to detect and avoid predatory mammals, or to find places where the temperature is appropriate for sleeping or sunbathing. Information from sensory dimples activates a region of the brain, the optic tectum, which usually processes visual input. Researchers believe that the brain then builds a three-dimensional image of the prey, similar to that obtained with a thermal camera, allowing the snakes to "see" in infrared.

Vampire bats, which, as their name suggests, feed on the blood of other creatures, are the only mammals known to have developed a fine ability to sense in infrared. These bats have an innervated epithelial pit, located on their nose. Infrared detection allows them to detect the heat of the blood flowing in the veins flush under the skin, and thus to choose the most appropriate area to bite their prey.

Invertebrates are not to be outdone. For example, Australian beetles of the genus *Melanophila* have infrared receptors on their thorax that enable them to detect forest fires from more than 100 kilometers away! Indeed, the adults of these species are attracted to forest fires, as their larvae can only develop on freshly burned wood. The adults mate in areas where the fire has just been extinguished, and the females lay their eggs under the bark of burned trees. After hatching, the larvae feed on the wood of fire-killed trees, taking advantage of an area deserted by predators. The detection system for these beetles is so effective that it is currently inspiring researchers to build forest fire detectors.

9.5. Information filtering

As we have seen in this chapter, among the almost infinite range of possible sensory information, animals primarily perceive the signals associated with their ecological needs through sensory organs that are the products of millions of years of selection. Indeed, recent research shows that the cost of maintaining sensory systems represents a significant fraction of the energy in animals, even outside the times when signals are perceptible. Indeed, sensory signals are perceived by modifying the resting potential of the receiving neurons. This resting potential, which is negative in relation to the environment outside the cell, is produced by ion pumps that continuously consume energy, even in the absence of a signal. The ability to detect *stimuli* from the outside world is therefore costly and evolution has favored trade-offs between this cost of sensory organs and their usefulness for the survival or reproduction of the animal. A first level of filtering of information is therefore carried out by the senses: for example, animals only perceive certain auditory frequencies or parts of the visual spectrum, and these limits correspond to what is relevant to them.

We have already mentioned the hearing of certain species of moths, which are predisposed to perceive the high-frequency sonar impulses of the bats that hunt them, and hardly perceive other sounds. The entire auditory investment of these moths, from the morphology of their auditory system to the physiology of auditory neural circuits, is devoted to the detection and analysis of this narrow ultrasonic frequency range.

Another major evolutionary pressure, reproduction, leads to the appearance of significant sexual dimorphisms in the peripheral sensory structures, as well as in the central circuits responsible for sensory processing. For example, it is easy to distinguish males from females in most species of moths because males have much larger antennae than females. These antennae, which are feather-shaped to provide a larger surface area, are covered with sensory cilia with extremely sensitive chemoreceptors, specific to the pheromones emitted by the females. Another example is the *polyphemus* butterfly, where males carry more than 60,000 senses per antenna, while tapered-antennae females only have 13,000. The male *Bombyx mori* can perceive the pheromones of females from a distance of 1 kilometer, but its olfactory receptors only respond to a narrow range of *stimuli*. In crickets, the neural circuits of females are specialized to distinguish the singing of male congeners from those produced by males of other species that are co-existing in the same habitat.

However, it is probably in the visual systems of insects that sensory filters are most noticeable. In some insects, such as mayflies, males have an extra pair of eyes for the sole purpose of spotting the passage of a female flying above them, appearing as a small dark silhouette moving rapidly and contrasting with the sky. The processing of information from this pair of eyes, both in the retina and in the optic lobe, is devoted solely to the detection and analysis of the movements of the females, and is particularly sensitive to contrast and movement information.

While sensory specialization is generally positive for its carriers, it is sometimes directed to the benefit of other species, which may exploit such limitations. One of the best-known examples involves bola spiders, which capture moths with a sticky silk ball (called a "bola" by analogy with the tool used by gauchos to capture cattle) suspended from the end of a silk lasso attached to the end of one of their front legs. To attract the moth, the spider secretes a chemical substance that mimics the composition of the pheromones emitted by the female moths in order to attract the males, who are thus victims of their own ultra-specialization in detecting this pheromone.

A similar exploitation of a visual filter is observed in the *Photuris versicolor* fireflies. In this species, males and females detect each other by emitting flashes of light through an organ located on the ventral side of the abdomen. The frequency of these emitted flashes is specific according to the species and sex. When a male spots a female, he flies towards her and the two individuals mate on the ground. However, after mating, the female becomes motivated by the search for food. She then changes her flash frequency to that of another species of firefly, *Photinus pyralis.* Her flashes attract a male *Photinus pyralis*, which the female *Photuris versicolor* eagerly devours.

Despite this specialization of the sense organs, the majority of animals are constantly receiving such a mass of information that they cannot process everything simultaneously. First and foremost, they have to be able to react to what is most important to them at any given moment. Another level of filtering thus consists of responding more to the most relevant aspects of *stimuli*, even if others are perceptible to the animal's senses. For example, the photoreceptors in the frog's retina are connected to form a receiver field. Some of these receptor fields respond more positively to small dark moving objects. When feeding, frogs focus on such objects, responding much less to all other *stimuli.* A second classic example in ethology is the male robin, attacking a lure consisting of a simple clump of red feathers. The robin is nevertheless capable, in other circumstances, of discriminating between a bird and a clump of feathers but, in a context of territorial defense, he focuses on the clump of red feathers.

In another classic experiment, Marian Dawkins (1961) trained chickens to feed on colored rice grains, placed on a substrate of the same or different color. She then observed that chickens that were fed grains that were clearly visible were more likely to spot grains that contrasted in color with the substrate, whereas chickens that were fed grains of the same color as the substrate were more sensitive to differences in texture or shape when looking for their food. Chickens are therefore capable of selective attention and pay more attention to characteristics that are habitually relevant in order to feed themselves.

In 1960, Luuk Tinbergen (the brother of the famous ethologist Nikolaas Tinbergen) observed the type of prey that great tits would bring to their young and compared his observations with the abundance of these prey, which varied according to the seasons. He observed that when the quantity of certain cryptic species (such as green caterpillars on green foliage) increases, the birds initially find very few and then suddenly start bringing back large quantities. Luuk Tinbergen proposes the concept of a "search image" to explain this phenomenon: cryptic prey can only be found in quantity once the bird starts looking for it. This observation was later confirmed in other bird species. It therefore seems that birds improve their ability to perceive cryptic prey after finding some; they are then able to picture the prey they are looking for and thus focus their attention on the characteristics of these prey. This "search image" is not necessarily limited to a visual representation: for example, tunas seem to form an "olfactory image" of certain prey, also in relation to their abundance.

9.6. Conclusion

Each organism inhabits a universe that represents the sum of all the information received and processed by the nervous system of that organism. The term *Umwelt* was coined, to describe this universe unique to each species, by Jakob von Uexküll (2010), who gives us the example of the tick. The tick remains motionless on a branch until it smells butyric acid (produced by the sebaceous follicles of all mammals), which causes it to let itself fall, and then it feels the warmth of the mammal on which it has fallen, which prompts it to use its tactile sense to look for an area with as little hair as possible where it will pump its host's blood. Thus, according to von Uexküll, the tick's *Umwelt* is essentially composed of three perception signs (butyric acid, heat and bare skin) and three effect marks (the behaviors aroused by these *stimuli*). The sensory universes of different species are also different, even if they live in the same environment: there are as many *Umwelts* as there are animal species.

It should also be noted that it is common for sensory systems to evolve with the development of an animal, meaning that the sensory universe in which the organism lives can often change over the course of its life. This is particularly striking in invertebrates, whose larval forms may be very different from adult forms. For example, a dragonfly spends its larval life, which lasts several years, underwater. Although it already has large functional eyes, its predominant senses are chemoreception (to locate its prey, but also its predators) and the sense of touch (to capture prey). After the metamorphosis, the adult dragonfly (which only lives for a few months) gains the ability to fly; it lacks tympanic organs, loses its chemoreception capacities and sight becomes the most important sense: the adult expresses more variants of coding genes for opsins than the larvae.

Some animals also have very specific sensory organs, defying the classification of sensory receptors: such as the famous twisted "horn" of male narwhals (nicknamed "sea unicorns" for this reason), which is in fact an incisor whose structure is the opposite of that of other vertebrates, since it consists of a core of hard tissue surrounded by a softer pulp and acts as a probe that transmits information to the brain about the temperature, pressure and salinity of the water.

Furthermore, even if we have preferred, for the sake of clarity, to present the different sensory modalities separately here, it is important to keep in mind that, in the majority of situations, an animal receives information from multiple sensory modalities at the same time, which it is able to integrate to form a complex representation of its environment. For example, it seems that bees, in their search for pollen or nectar, first rely on a genetically predetermined olfactory "search image", but this image is then, through experience, enriched into a multimodal search image including the color of the flower, its form, etc.

In this chapter, we have seen that sensory universes inhabited by other animals can be very different from the world we perceive. Thus, in his famous article "What It Is Like to Be a Bat", the philosopher Nagel (1974) wonders whether we can really know what it is like to be a bat and answers in a negative manner: according to him, even if a human tries to imagine what it would be like to have wings, to find his way around thanks to echolocation, to spend the day suspended by a foot in an attic, etc., it is not possible to know what it would be like to be a bat; it would only be a human mental projection, which would not allow him to understand how a bat really feels.

Therefore, we can't put ourselves in the place of a bat or any other species. However, there are overlaps between the sensory systems of different species, making interspecific interactions and the study of other species by researchers possible. Advances in technology also allow us to perceive *stimuli* that are beyond human sensitivity. We don't know what it feels like to use biological sonar, but knowing that bats have one makes it easier to understand their behavior and ecology.

Recent research shows that the interaction of sensory systems with the environment can be an important factor in the evolution of species and that it could favor the appearance of new species, without there necessarily being any geographical separation: for example, the visual system of certain fish could influence the preferences of females for different colors of males; researchers have also shown that echolocation signals can diverge within the same bat species, a divergence which could give rise to speciation.

By studying the different types of *stimuli* perceived and used by animals, we can better comprehend their sensory universes, and better understand the complex links and reciprocal influences between the sensory systems, evolution and ecology of these organisms.

9.7. References

Antinucci, M., Risso, D. (2017). A matter of taste: Lineage-specific loss of function of taste receptor genes in vertebrates. *Frontiers in Molecular. Biosciences*, 4(81).

Burnett, S. (2011). Perceptual worlds and sensory ecology. *Nature Education Knowledge*, 3(10), 75.

Campbell, N., Reece, J., Urry, L., Cain, M., Wasserman, S., Minorsky, P., Robert, J. (2012). *Biologie*. Pearson, London.

Dawkins, M. (1961). Perceptual changes in chicks: Another look at the "search image" concept. *Animal Behaviour*, 19(3), 566–574.

Gagno, S. (2013). L'audition chez les chéloniens : Cas de la Tortue d'Hermann. *Nature de Provence – Revue du CEN PACA*, 2, 91–98.

Land, M. (2019). Senses [Online]. In *Encyclopaedia Brittanica*. Available at: https ://www.britannica.com/science/senses [Accessed February 3, 2019].

Marshall, J., Arikawa, K. (2014). Unconventional colour vision. *Current Biology*, 24(24), R1150–4.

McFarland, D. (2009). *Le comportement animal : Psychobiologie, éthologie et évolution*. De Boeck Supérieur, Louvain-la-Neuve.

Morley, E.L., Jones, G., Radford, A.N. (2014). The importance of invertebrates when considering the impacts of anthropogenic noise. *Proceedings of the Royal Society B: Biological Sciences*. 281.

Nagel, T., (1974). What is it like to be a bat. *Philosophical Review*, 83(4), 435–450.

Tinbergen, L. (1960). The natural control of insects in pine woods. I. Factors influencing the intensity of predation by song birds. *Archives Néerlandaises de Zoologie*, 13, 265–343.

von Uexküll, J. (2010). *Milieu animal et milieu humain*. Rivages, Paris.

Warrant, E.J. (2016). Sensory matched filters. *Current Biology*, 26(20).

10

Emotional and Social Forms of Robots

Laurence DEVILLERS
LIMSI-CNRS, Sorbonne University, Paris, France

10.1. Introduction

The emotional and social forms one might have thought to be characteristic of humans are invading machines, which are closer to human capacities. For a long time, philosophers and scientists have pitted reason against emotion. Today, the evolution of scientific knowledge through neuroscience shows that emotional forms are necessary for cognitive functioning, memorization and learning. *Empathy*, for example, is one of the pillars of human relationships and relevance. It is an emotional response to a particular situation. Being able to feel an appropriate emotion in response to that expressed by another, distinguishing between the emotion of the other (*your child is in pain*) and your own emotion (*you are in pain for him, but you do not feel physical pain*), is a form of thinking that uses memory (Jaffard 2019)[1] and sensoriality (Bovet 2019)[2]. Humans possess a particular ability to project themselves. In children, affective empathy appears at the age of one, cognitive empathy appears later at the age of around four and a half years. An empathy including the sense of reciprocity and allowing the construction of a sense of morality and justice develops between the ages of 8 and 12.

1 Chapter in this book.
2 Chapter in this book.

The Explosion of Life Forms,
coordinated by Georges CHAPOUTHIER and Marie-Christine MAUREL. © ISTE Ltd 2020.

In order to understand the particularities of the emotional and social forms of robots, let's already be reminded that a robot is characterized by three interacting components: it collects data through its sensors; it interprets them through its programs; it can move and act on its environment. In addition, it can have an anthropomorphic appearance and social language interaction capacities. Service robotics tend to create social and affective robots through the ability to integrate emotional models in the design of dialogue systems. A chatbot or "conversational agent" is also referred to as the human-machine dialogue system. Emotional and social capacities are obtained through statistical models driven by data, whether it be through perception or generation. These systems do not use formal models but learning approaches such as deep learning. Deep learning is a set of machine learning methods attempting to model with a high level of data abstraction, through the articulated architectures of different non-linear transformations. Robots, chatbots like Google Home or Siri on our phone can give us the illusion that machines are empathetic if they are given the ability to identify the emotions of their human individual, which would be based on detected emotions and generated through facial expressiveness, gestures, postures and acoustic expressiveness of emotions. Detecting a person's emotions can cause the machine to change response strategies. This allows it to respond to someone when it has detected sadness with "I'm sad too". This deception may lead us to believe in the "empathy" of robots. But this is not really "empathy" because these machines do not have phenomenal consciousness, that is to say, experiences that characterize a person's "experience" or "feeling". Phenomenal consciousness, unlike consciousness defined as "cognition", is associated with a qualitative experience, such as a feeling of pleasure or pain, hot or cold, and so on. Phenomenal consciousness is therefore not reducible to the physical or physiological conditions of its appearance, and it is inseparable from the subjectivity of the individual. There is thus a great difference between scientific descriptions of consciousness, which refer to publicly observable behavior or brain functioning, and the subject's own phenomenal consciousness.

The neuroscientist Antonio Damasio (1994) brings a new vision of how emotions manifest themselves, in the close interrelationship between the body and the brain in the perception of objects. The notion of the "body" is central: man cannot be reduced to a thought, to a self-awareness; he is also a body, through which he finds himself in the world. In his major work *Phenomenology of Perception*, the philosopher Merleau-Ponty (1945) gives the body the importance it had lost in the philosophical tradition inherited from Descartes.

However, we have to note that the reduction of thought to brain activity or the reduction of man to machine are current scientific topics. Some transhumanist authors, such as Nick Bostrom (2014), put forward the hypothesis that the phenomena that we call "thought" and attribute to the brain are independent of its support and that there is, therefore, a complete permeability between the brain and its digital simulations, which can quickly exceed our own capacities. Without adhering to these hypotheses, which are currently untestable, the impact of reciprocal adaptation situations deserves to be studied, all the more so as the situations blur the perceptible boundary between that which is human and that which is artificial.

Trying to find the link between objectivity and subjectivity is fundamental and is clearly lacking in the current understanding of our projection of machines. Scientific demystification is urgently needed. The veils of mystery of sciences such as artificial intelligence, cognitive neuroscience and "affective" intelligence must be lifted in order to make better use of these artifacts, robots or social and affective conversational agents in society. Computer modeling of affects raises the question of the societal consequences of living in a daily life surrounded by pseudo-affective objects (Devillers 2017). Such an example is the possible ethical principles imposing a distinction between artificial agents.

10.2. Communication with social and emotional robots

The field of affective computing has its roots in the 1997 work of Rosalind Picard (1997) at the Massachusetts Institute of Technology (MIT), and brings together three technologies: the recognition of human emotions, reasoning and decision-making using this information, and the generation of emotional expressions. This field is inherently multidisciplinary. The recognition of social messages conveyed by faces and voices, particularly emotional expressions, is an indispensable element in human communication and the implementation into all societies. In a consensual way, emotion is defined as a reaction to a real or imaginary event/situation, comprising several facets or components. Three components are generally accepted as essential constituents of the emotional response. These are subjective feeling (emotional experience), physiological reaction and emotional expression (facial, vocal or postural). In a context of communication, expressions are transformed according to a set of social-cultural rules. These rules vary from one culture or social group to another, in "objectively" similar contexts. Some social situations may require the suppression of certain expressions, while other situations, on the contrary, may require showing specific expressions, or even exaggerating them. Hiding a spontaneous expression that would be undesirable in a given social context is also possible. The social context may influence emotional expressions, depending on the issuer's position and objectives in the situation. In a communication situation, an individual may use his emotional expressions in such a way as to influence – more or less (un)consciously and more or less (in)voluntarily – the reactions of his correspondent.

Machines will increasingly interact vocally with us in everyday life. Conversational agents and emotional and social robots can already enter systems for detecting, reasoning and generating emotional expressions that, even with significant errors, can interact with us. They are now invading our private spheres. In the United States, there are up to six Alexa or Google Home speakers per household, one per room. The market for "isolated" people is huge: these machines could accompany them on a daily basis, to monitor their health, educate them, help them and entertain them, in short, take care of them! For these tasks, the machine is created to be a kind of digital companion, assistant or supervisor.

"Informing is not communicating. Communicating is as much about sharing what we have in common as it is about managing the differences that separate us," explains Dominique Wolton (2009). Communicating with machines is, above all, an exchange of information and not of meaning, with uncommunication happening too often. Miscommunication can occur even in circumstances where information is communicated, for example, if the information does not contain a message or if the receiver, such as a robot, cannot decode the information contained in the message. Machines are far from having sufficient semantic capacity to converse and share ideas, but they will soon be able to detect our discomfort, our stress, perhaps even some of our lies.

10.3. Human empathy for machines

Reeves and Nass's *Media Equation* (Reeves and Nass 1996) tells us that we apply the same social expectations as with humans when we communicate with artificial entities, and unconsciously assign rules for social interaction with them. Anthropomorphism is the attribution of behavioral or morphological characteristics of human life to objects. With this reflex, both innate and socially reinforced, an object that seems to be in pain can inspire empathy. Experimental studies have shown the projection of affective and probably empathic reactions towards artificial entities, such as deliberately damaged toy robots.

Researchers found that humans empathize with abused robots, albeit to a lesser degree than with abused humans, but this empathy does not exist with inanimate objects. Using brain imaging, recent research indicates that individuals respond to emotional images of humans and artificial entities in surprisingly similar ways. If we represent artificial entities as humans, then perhaps it is not surprising that we respond emotionally to artificial agents in the same way that we do to humans.

However, it is not certain that representations of the same complexity can be attributed to robots.

The robots Nao, Pepper and Romeo[3] are social robots that could also be mirrors of our emotions, but they are not strictly speaking empathetic. Fan Hui, the Go Champion who trained AlphaGo[4] for Google Deepmind, explained that playing against a machine was a bit like playing against yourself, because you projected your emotions onto the machine, which reflected them like a mirror. The *Paro* seal robot, developed in Japan in 1993, was marketed in Japan in 2005, then in the United States in 2009 (with FDA certification as a therapeutic robot). When Paro is used in EHPAD with elderly people, the robot's reactions are not "empathetic"; they are more like expressive pet behaviors. During our tests in EHPAD (Garcia *et al.* 2017) with the robots Nao and Pepper, which detected emotions and adapted accordingly, the elderly people we met (a good 50 people with an average age of 85 years, who were not under guardianship) had reactions of curiosity and amusement. However, they did not believe that the robot "understood" them. Instead, they said that it could detect some of their behaviors.

A two-year-old child knows that his cuddly toy is not alive and yet he talks to it as if it were alive. Psychologists talk about dolls and cuddly toys as transitional objects. The transitional object is therefore a privileged object, chosen by the child, generally soft to touch. It allows the baby to fight against anxiety. It is, in a way, "his/her" first possession. It is neither perceived as part of the mother, nor as an interior object. It allows the child to move from the subjective to the objective. Paro, the stuffed robot, can be seen as a transitional object.

Social robots are also primarily technical objects that store our data in order to transfer it to a doctor, for example. Life-supporting robots are also part of an ecosystem that includes many actors: family, caregivers, doctors.

Behavioral problems with robots are studied in psychology and psychiatry, notably by Serge Tisseron (2015). Kate Darling (2016), a researcher at the MIT Media Lab, is studying people's empathic reactions to robots, particularly in cases of robot abuse, with the idea that robots should be given legal protection just like animals have. However, animals are living organisms that can feel emotions and suffer, which is absolutely not the case with robots, which only simulate emotions.

---

3 Robots from Softbank Robotics company.

4 AlphaGo is a computer that won the game of Go against Lee Sédol, World Go Champion.

10.4. Machine emotions

Giving machines the ability to interpret and simulate emotions is essential for building systems that are capable of social interaction and better communication with humans. There are many applications to help people who are dependent, such as those in old age or suffering from various degenerative pathologies. In some cases, a hesitation, a *breath* from the machine gives the impression that the machine is "alive". The American philosopher Donal Davinson describes anomic monism as the union expressed by two different languages without translation. There is no causal reason between mind and body. The more fragile the machine looks, the more one can humanize it and be moved by it, even if strictly speaking it does not simulate empathic behavior.

In Spinoza's philosophy, and more particularly in his book *Ethics* (Spinoza 1677), the biologist and philosopher Henri Atlan sought to find a source of inspiration to explain today's world and the relationship between body and mind. Spinoza explains that the organism makes itself. Affects and hatreds, the ideal places where body and mind are united, are sources of alienation if we undergo them or of freedom if we understand the underlying mechanisms. Williams James, with his theory of emotions in psychophysical science, is a precursor of today's cognitive neuroscience. Through the exploration of the brain, we can prove what he was saying, which was contrary to common sense. Body expressions precede feeling. Body and mind are mixed, Spinoza teaches us. However, the machines do not yet have bodies in terms of viscera, hormones and "skin"; they have no intention, pleasure and desire of their own, that is to say, no *conatus* in Spinoza's sense. Lastly, the living can be defined as autonomous and have the ability to reproduce themselves.

At present, the relative autonomy of robots is still programmed by humans. The programmed learning ability can offer more or less freedom to the machine. Giving a robot the ability to learn on its own, in interaction with the environment and humans, is the Holy Grail for researchers in artificial intelligence. If robots learn on their own, it would be desirable to control what they learn. However, this ability constitutes a technological and legal breakthrough, and raises many ethical questions. These robots can be, in a way, creative and autonomous in their decision-making, if they are programmed to do so.

To want to copy man's intelligence onto a machine is very narcissistic, because what do we know about our intelligence? We do not know the substratum of thought: we are only aware of a small part of our perceptions and brain activity. There is no term more polysemic and subject to interpretation than "consciousness":

for some people, it evokes self-consciousness, for others, consciousness of one's neighbor, or even phenomenal consciousness, moral consciousness, etc.

With a materialistic philosophical conception of life, we can consider the computer and the human brain as comparable systems, capable of manipulating information. The most efficient digital modeling, such as *deep learning*, is based on a simplified modeling of the neuron (formal neuron) embedded in a computer-simulated state machine. The number of hidden layers in the model architecture corresponds to the depth. For the moment, we are very far from the complexity of the living!

Experiments conducted at Neurospin by Stanislas Dehaene's team (Dehaene 2014), using subliminal images, have shown that our brain mainly functions in an unconscious mode. Routine actions such as recognizing faces and words, for example, are carried out without recourse to consciousness. In order to access consciousness, the human brain implements two types of information processing: a first level, called "global availability", which corresponds to the vast repertoire of information – modular programs that can be called up at any time to use it, and a second type of information processing associated with it, specific to human consciousness, called self-monitoring or self-assessment, "the ability to process information about oneself, which can also be called metacognition". Thus, the brain is capable of introspection, of controlling its own processes and obtaining information about itself, which leads to autonomy.

Today's artificial intelligence systems have the ability to calculate correlations of facts. For example, with deep learning approaches, to make decisions and to learn, but without being aware of them. Some robot prototypes, however, already have embryos with levels of "consciousness" comparable to those described by Stanislas Dehaene (2014). They are simulated by mechanisms of knowledge sharing and introspection. For all that, these machines are not conscious like a human being can be: they have neither moral consciousness nor phenomenal consciousness associated with a qualitative experience, such as the sensations of hot or cold, a sense of anxiety, etc., because they have no viscera or feelings, unless, once again, they are simulated.

The first learning and communicating robots with pain and pleasure sensors (Asada 2015) interact using simple processes: sensors on the robot's body, a camera and a microphone that allow them to associate a face and a voice with the expressive signs they receive. Their expression indicators (emitted sounds, diodes) allow humans to understand their states. To produce rapid effects, physical actions are required: a stroke on the head and they associate a positive criterion with the person they see; a slap on the head and they associate a negative criterion. These machines learn and adapt their behavior to the contexts in which they find themselves. However, interpretations are still very limited. For example, a robot assistant to dental surgery students should give the best information on the faces of pain and anticipate gestures to give alerts on the proximity of a nerve and possible suffering. Professor Minoru Asada, from the University of Osaka (Asada 2019), believes that robots should get as close as possible to humans, so that interactions are as transparent as possible and learning can take place by imitation. But an artificial consciousness, endowed with feelings, thoughts and free will without human programming, is still unlikely to emerge spontaneously with current computer architectures.

10.5. Conclusion: risks and benefits

Daily life with these companion robots could lead to long-term emotional and social risks, which must be anticipated in order to benefit from these machines. Addiction to the machine and the isolation of humans, as well as the transfer of autonomy to the machine, the confusion between the machine and the human, are deviances that need to be addressed. One of the risks, especially for fragile people, is to forget that a robot is connected and programmed. The ability of a robot to adapt to its owner could be used to make him/her accomplish certain choices rather than others, notably to help him/her better manage his/her deviant behaviors (sexual pathology, drug addiction), but also for less laudable causes in the field of consumption. Another risk is forgetting that a robot does not feel anything, has no emotions, has no consciousness and is not alive. While it is possible to feel empathy for a robot and talk about a robot's suffering, it is important that elderly people, who may put their lives in danger to help their robot, realize that a robot does not suffer, even if it falls; they must be aware that it is just a programmed object.

The emotional and social forms of robots raise many ethical, legal and social questions. Among these questions, we can ask ourselves:

– who is responsible in case of an accident: the manufacturer, the buyer, the therapist or the user?

– how do you regulate their functioning? Should moral rules be integrated into their programming? Should their use be controlled by permits?

– how will our privacy and personal data be preserved?

– every system must be evaluated before being put in the hands of its user. Can we prove that it will be confined to the functions for which it was designed, that it will not exceed the limits set?

– the data that the machine uses for its learning directs it towards certain actions. Who will supervise the selection of this data?

– what tasks do we want to create these empathetic artificial forms for?

These important issues have only recently come up. Spectacular advances in digital technology will one day make it possible to improve people's well-being, provided that we don't think about what we can do with it, but about what we want to do with it. A number of ethical values are important: the ethics and responsibility of designers, the emancipation of users, evaluation, transparency, explicability, fairness and equity of systems, and the long-term study of human-machine co-adaptation (the machine will adapt to the human and the human to the machine). Control by humans will always be paramount. It is necessary to develop ethical frameworks for emotional and social robots, especially in health, and to understand the level of human-machine synergy.

The artificial intelligence research community is beginning to worry (Rahwan *et al.* 2019): we design machines in our own image and must anticipate the consequences for humans and society. The emotional and social forms of robots are certainly only simulations, but if machines adapt to us, we will also adapt to them. My book *Des robots et des hommes: myths, fantasies et réalité* (*Robots and Men: Myths, Fantasies and Reality*) (Devillers 2017) is a work to give warning about the lack of regulation and ethical standards on the design of these machines, and the lack of long-term research on this "human-machine co-evolution".

We need to demystify, train all citizens in artificial intelligence and put human values back at the center of the design of these robotic systems.

10.6. References

Asada, M. (2015). Towards artificial empathy. *International Journal of Social Robotics*, 7, 19–33.

Asada, M. (2019). Artificial pain: empathy, morality, and ethics as a developmental process of consciousness. *Philosophies 2019*, 4, 38.

Atlan, H. (2018). *Cours de philosophie biologique et cognitiviste : Spinoza et la biologie actuelle*. Odile Jacob, Paris.

Bostrom, N. (2014). *Superintelligence : chemins, dangers, stratégies*. Oxford University Press, Oxford.

Damasio, D. (1994). *L'erreur de Descartes*. Odile Jacob, Paris.

Darling, K. (2016). Faut-il accorder une protection juridique aux robots de compagnie. In *En compagnie des robots*, Bensoussan, A., Constantinides, Y., Darling, K., Ganascia, J.-G., Tesquet, O. (eds). Premier parallèle, Paris.

Dehaene, S. (2014). *Le code de la conscience*. Odile Jacob, Paris.

Devillers, L. (2017). *Des robots et des hommes : mythes, fantasmes et réalité*. Plon, Paris.

Dumouchel, P., Damiano, L. (2016). *Vivre avec les robots*. Le Seuil, Paris.

Garcia, M., Béchade, L., Dubuisson Duplessis, G., Pittaro, G., Devillers, L. (2017). Towards metrics of evaluation of Pepper robot as a social companion for elderly people. *International Workshop on Spoken Dialogue Systems (IWSDS)*, 8.

Merleau-Ponty, M. (1945). *Phénoménologie de la perception*. Gallimard, Paris.

Picard, R. (1997). *Affective Computing*. MIT Press, Cambridge.

Spinoza, B. (1677). *L'Éthique*. Livre III, Proposition II. et SCOLIE. Points-Essai, Paris.

Rahwan, I., Cebrian, M., Obradovich, N., Bongard, J., Bonnefon, J.F., Breazeal, C., Jennings, N.R. (2019). Machine behaviour. *Nature*, 568(7753), 477.

Reeves, B., Nass, C. (1996). *The Media Equation*. Stanford University, Stanford.

Tisseron, S. (2015). *Le jour où mon robot m'aimera : vers l'empathie artificielle*. Albin Michel, Paris.

Wolton, D. (2009). *Informer n'est pas communiquer*. CNRS, Paris.

11

When Medical Technology Mimics Living Forms

Didier FASS
LORIA-CNRS, University of Lorraine, Nancy, France

11.1. Introduction

Medical technology is the practice of conceptualization and design of technical objects and medical devices.

The complete or partial loss of a limb or the alteration or destruction of an organ or tissue of the body diminishes its capacity, functional stability or viability, be it temporarily or permanently. In order to compensate or repair, both temporarily or permanently, these losses or failures, medical engineering, which relies on art, craftsmanship and, today, on science and technology, has designed and manufactured many artificial systems for repairing, supplementing and extending the field of the body and life. Located at the juncture of biology and medicine, engineering sciences and human and social sciences, in addition to being in the movement of a pluridisciplinary convergence of nanotechnology, biotechnology, computer and cognitive sciences (Rocco and Bainbridge 2002), many medical technologies are inspired by or mimic the forms of life (structural or anatomical forms) and living forms (dynamic or physiological forms).

Today, these technologies are the foundation of the fields of repaired and augmented humans. They are the subject of pluridisciplinary research and

The Explosion of Life Forms,
coordinated by Georges CHAPOUTHIER and Marie-Christine MAUREL. © ISTE Ltd 2020.

development, and of philosophical and political debates on their ethical, economic and strategic stakes. Their theme feeds a number of catastrophic or all-powerful fantasies. In terms of their epistemological and technological foundations[1], these technical objects thus question our ideology and our conceptualization of life, of living forms, of human beings and their place in nature. Because they are polysemic and pluridisciplinary, the critical analysis of these concepts establishes our relevant conception of current and future medical technology.

Mimicking the living, and its forms, to develop and design medical technologies (orthoses, prostheses and other systems of electro-physiological or cognitive functional substitution, both intra or extracorporeal) is a necessity to reconstruct anatomy and restore physiology. For more than 3,000 years, from the earliest dentures or toes to contemporary cardiac prostheses, the matter of mimicking the morphological or functional form of an altered or lost body part has been essential. After addressing the question of imitating the living and form in a historical and epistemological perspective, we will extend critical reflection to forms of thought or machine processing – both medical and symbiotic. We will explain the associated concepts and the theoretical principles necessary to understand the organization of living organisms, in order to design and model medical artifacts in integrative bioengineering. We will thus describe the basic needs for biocompatibility and biointegration, and define the fundamental requirements, including form(s), of "human-artifice" engineering for medicine and for the "enhanced human". Lastly, we will put forward a perspective on integrative bioengineering for the bioprinting of implantable medical devices and tissue engineering.

11.2. Historical and epistemological perspective

The conceptualization, perception and representation of nature, particularly of living forms, is essential to the design of medical technology. It is set in a historical and epistemological perspective.

11.2.1. *A comparative history of medical technology*

Generally speaking, technology – techniques, processes and methods that draw on or mimic living forms and systems – are said to be "bio-inspired". They

1 The word "technology" and the adjective "technological" are used here in the sense of the foundations of scientific knowledge of technology and the principles of the constitution of technical objects.

constitute the field of biomimetics[2] (Bar-Cohen 2006) and bionics[3]. This approach can be found in artificial intelligence[4], cobotics[5] or synthetic biology[6].

From passive prostheses (teeth and limbs) to active prostheses, neuroprostheses and body or organ-machine interfaces, from tissue bioengineering to 4D bioprinting of tissues and organs, the history of medical technology has been built up over many centuries in the invention and evolution of medical devices[7].

---

2 The word "biomimetics" was coined by Otto H. Schmitt (Schmidt 1969). It is a pluridisciplinary field of which the objective is to derive general scientific and technological concepts and principles, in particular living forms, from the observation of nature, for the design and construction of technical systems. The main current fields of application of biomimetics are engineering, architecture, medicine and art (Hwang *et al.* 2015), without forgetting space, which is the fertile ground for this new technology.

3 The word "bionics" refers to the science of systems that have a function copied from nature or that represent the characteristics of natural systems or their electronic analogues (Vincent 2001). It was coined in 1958 by Jack E. Steele, a US Air Force neurospychiatrist, whose research topic was the "cyborg". The latter term, coined by Manfred E. Clynes and Nathan S. Kline (Clynes and Kline 1960), refers to "devices necessary for the creation of self-regulating man-machine systems that deliberately incorporate exogenous components that expand the body's self-regulating control function, particularly its homeostasis, in order to adapt it, unconsciously, to new environments".

4 The term "artificial intelligence" was coined in 1955 by McCarthy, Marvin Minsky, Allen Newell and Herbert A. Simon (McCarthy *et al.* 1955) for the organization of a prospective summer school in Dartmouth "on the basis of the conjecture that every aspect of learning or other characteristic of intelligence can, in principle, be described with such precision that a machine can be designed to simulate it".

5 The word "cobot" was defined in 1996 by Ed Colgate, Witaya Wannasuphoprasit, Michael Peskin (Colgate *et al.* 1996) "as a robotic device that manipulates objects in collaboration with a human operator. It provides assistance to the human operator by configuring virtual surfaces that can be used to constrain and guide motion". By extension, the word "cobotics" refers to the field of direct human-robot collaboration and interaction in the performance of an activity, such as that implemented in today's surgical robots or modern human-piloted aircraft.

6 Synthetic biology is an emerging interdisciplinary field comprising many specialties from biology, engineering (especially systems engineering), physics, chemistry, computer science, epistemology, etc. While it initially focused on genomic engineering and systems biology (Cameron *et al.* 2014), today it promises a fundamentally deeper understanding of how living systems function, in addition to the ability to simulate and recreate them for medicine, public health and the environment, including renewable energy (NAKFI 2010).

7 Medical devices are defined in Articles L5211-1 and R5211-1 of the French Public Health Code as "any instrument, apparatus, equipment, material, product, with the exception of products of human origin, or other article used alone or in combination, including accessories

As early as the first millennium BCE in ancient Egypt, the first dental or toe prostheses, found on mummies or in sarcophagi, were carved (Finch 2011). There are traces of foot prostheses dating from the early Middle Ages, before the spread of pestle and hooks in the Renaissance. It was in the 16th Century, with the descriptive and formal anatomical works on the structure of the human body by Andreas Vésale[8] and the influence of Ambroise Paré, that medical technologies appeared. Based on this early work on systemic anatomy and his practice of battlefield surgery, Ambroise Paré designed and drew a number of surgical instruments for the repair of bleeding wounds, head injuries and amputations. Thus, confronted with new types of injuries due to the appearance of powder weapons, he designed fixed prostheses for teeth, palates[9], legs or hands, and thought up articulated mechanical prostheses for hands, arms or legs (Paré 1585) based on their morphology and osteoarticular system. For dental prostheses, examples from as early as 600 BCE also exist among the Etruscans, and it was with Pierre Fauchard (1728) that dentistry really converged with technology to give conservative and prosthetic care.

The invention of the microscope at the end of the 16th Century marked a decisive moment in the description and understanding of anatomy, and its relationship to physiology. It allowed Marcello Malpighi[10] to discover the pulmonary capillary circulatory system that had eluded William Harvey[11] in his direct observations and precise description of the circulatory system. Until then, the arteries and the left side of the heart had been empty of blood on corpses, and they had been assigned the role of circulating the breath of life or soul through the body. Malpighi's discovery made it possible to conceptualize the body as a set of functional elements connected at different scale levels.

---

and software necessary for the proper functioning thereof, intended by the manufacturer to be used in humans for medical purposes and whose principal intended action is not obtained by pharmacological or immunological means or by metabolism, but whose function may be assisted by such means. Software intended by the manufacturer to be used specifically for diagnostic or therapeutic purposes also constitutes a medical device. Medical devices which are designed to be implanted in whole or in part in the human body or placed in a natural orifice, and which depend on a source of electrical energy or on any source of energy other than that generated directly by the human body or gravity for their proper functioning, are referred to as 'active implantable medical devices'".

8 *De humani corporis fabrica*, published in 1555.

9 Palatal obturator.

10 *De pulmonibus observationes anatomicae*, published in 1661.

11 *Exercitatio Anatomica de Motu Cordis et Sanguinis in Animalibus*, published in 1628.

In 1800, Marie François Xavier Bichat[12] founded the concept of tissue and established the link between the degradation of functions and tissue alteration in organs. He also describes how this phenomenon spreads in the organs, and in the space and time of the dying body. He thus establishes the essential role of tissues as fundamental anatomical elements, the cause of the physiological properties and pathological transformations of the organism. In 1848, some time before the work of Joseph Lister on antisepsis and the work of Louis Pasteur on asepsis and vaccination, which scientifically confirmed the theory of microbial diseases, Ignaz Semmelweis demonstrated the causal role of habits and organization of work (according to a modern formulation) on post-partum mortality and the importance of hygiene (hand and instrument washing) by his comparative observations between two obstetric clinics.

The end of the 19th Century marked the entry of medicine and medical technology into the modern era. Claude Bernard (1984) developed a general physiology on the experimental method, while doing so as a theorist. He envisaged the mathematical intelligibility of biological nature but, noting the enormous difficulties of a formalized representation, he preferred the discovery of new facts through experience (Chauvet 1995). He is thus at the origin of "evidence-based"[13] medicine and medical technology.

With the use of new weapons causing new forms of injury in the American Civil War, and particularly in the First World War, reconstructive and restorative surgery techniques as well as prostheses, especially maxillofacial prostheses for "broken faces", developed (Destruhaut 2011). The inter-war period was marked by the discovery of the antibiotic phenomenon and penicillin by Alexander Fleming in 1928. Its industrial production was accelerated with the Second World War. With the Second World War, knowledge of life forms and medical technology progressed, sometimes without any rules or ethics, and technological progress came from a therapeutic necessity for revival or rehabilitation techniques and physical medicine on the battlefield.

The need for new means of detection and armament encouraged the gathering of personalities with pluridisciplinary scientific and technical skills. This had a major effect on the evolution of concepts in medical technology, and on the way in which living forms were mimicked, or rather, modeled. Mathematician Norbert Wiener worked with physiologist and physicist Arturo Rosenblueth on the development of automated anti-aircraft ballistic tracking systems. They drew an analogy between

12 *Traité des membranes* et *Recherches physiologiques sur la vie et la mort*, published in 1800.

13 Guyat *et al.* 1992.

motor control by the nervous system and the automatic closed-loop control systems of the anti-aircraft battery tracking system. They formulated epistemological consequences on the role of models in science (Rosenblueth and Wiener 1945) and opened the field of "model-based" engineering. Norbert Wiener developed a control theory for the animal and the machine: cybernetics (Wiener 1948). Alan Turing's (1950) abstract separation of the "universal" machine from the program and John von Neumann's (1958) analogy between the computer and the brain strongly influenced interpretive concepts of biology, especially molecular biology. In the late 1950s, formal thinking, or cognition, sacralized a machine-program and body-brain dualism. It complemented the reductionist physicalist and functionalist paradigm, itself influenced by a metaphorical interpretation of cybernetics, the mathematical theory of communication (Shannon 1948), and emerging computer science and artificial intelligence. Combined with mechanical physics as a constitutive principle of machines and electronics, this had a major influence on the modeling of the living, and the abstract and formal representation of living forms and behavior.

Today, in the so-called "transhumanist" project, some people fantasize about concretizing the singularity of the living in a convergence with almighty computing. For them, the body is just a programmable or reprogrammable universal machine, whose operating system would have DNA as its core and the brain as its syntactic, symbolic or semantic interpreter[14].

This metaphysical position requires an epistemological revision, a theoretical renewal based on the concepts of the living form and the scientific principles of integrative physiology. These are the conditions for the formulation of a modeling framework for integrative bioengineering of medical technology.

In recent years, work on cardiac prostheses or vascular implants and membranes for dialysis filtration or extracorporeal blood oxygenation, and their associated inflammatory, immune or thromboembolic risks, have shown the need to rethink biocompatibility and biointegration. To reduce haemostatic risks, the whole Carmat® heart prosthesis uses unspecified bovine pericardium membranes for all the parts in contact with the wearer's blood. The Zurich Heart®, another whole-heart project, is working on the use and cultivation or autografting of endothelial cells, the cells that line our blood vessels, on the inside of these circulatory elements.

Tissue bioengineering and regenerative medicine have so far been based on biological and medical knowledge gained from experimentation. With the advent of prototyping, modeling and digital simulation techniques, in addition to the possibility of producing implantable bioartificial objects by printing, a new approach

14 Which is absolutely not where we stand.

based on models of living systems needs to be considered, revising conceptual representations of the nature of living forms and thus founding correct design ontologies of artificial living forms for medicine.

Since the 1950s, manned spaceflight programs have stimulated new concepts and solutions by understanding the nature of life, as well as the importance of the environment in maintaining the astronauts' bodies in shape[15], necessary to preserve their health and maintain them in operational condition according to the identified risks[16].

Beyond the mechanics, the physical-chemical properties of biomaterials and the logical-mathematical properties of living things, a change in the perspective of living things (Damish 1987), or in the representation of life and living forms, is necessary. Just as art has emancipated itself from the imitation of nature and questions new relationships to the body or ideas through concrete or abstract, virtual or actual plastic forms, integrative bioengineering and its integrative organizing principles, as we consider them in the following section, allows us to model bioinspired medical technologies and bioartificial objects in their possible structural and dynamic forms and evolutions, but in a humanly acceptable manner.

11.2.2. *Epistemological perspective*

The "human condition", and of his humanity[17], is his capacity to conceive and manipulate an artifice in a reflexive consciousness and imagination of the individual or social body that is located in an uncertain environment, with the desire to make it controllable, predictable and safe.

Compared to other technologies from the crafts and engineering sciences, the specificity of medical technology[18], the medical or dental arts, bioengineering and biotechnology, is that it directly affects the biological singularity and our individual and collective human future. Our biological singularity, although existing in physical space, is not reducible to its physical-chemical properties (Bernard 1858; Chauvet 1995) nor to its physical-mathematical conceptualization (Bailly and Longo 2006).

15 In the literal sense of the word.

16 Available at: https://humanresearchroadmap.nasa.gov/.

17 A set of characteristics specific to the human species in relation to other living species.

18 As an activity of conceptualization and design of medical objects and devices.

As Konrad Lorenz (1984) points out, "the processes of living things are quite different from mere physical-chemical processes". Although eukaryotic or prokaryotic, single-cell or multi-cellular living forms are inherited or determined by phylogenesis for their archetypal species-related morphological characteristics (Prochiantz 2008); they are modulated by epigenesis in interactions with the environment[19] (Heard 2013) in a non-deterministic or non-mechanistic way during ontogenesis, for adaptation, evolution and individual existence, from that which is normal to pathological, and from life to death.

That is to say, there is no calculable or non-calculable linear mechanical causality in a complete space of probable discrete states (Longo 2010; Lechermeier 2015). This ontological irreducibility, and the non-deterministic character of living systems, result in two fundamental problems for the sciences and conception methods of medical technologies for treatment, replacement, rehabilitation or enhancement:

– the coupling of systems of a different nature: the fundamental problem of the conception of human-machine systems, in particular those directly related to bodies such as medical technologies for repaired or enhanced humans, is that of the functional or *physiological* coupling of two systems of different natures or constitutions: on the one hand, the body, which is of a biological nature, and on the other hand, artificial systems (cyber-physics), which is of a physical and/or logical nature (Fass 2012). For biomaterials and implantable medical technologies, the chemical nature of the objects produced must also be taken into consideration for their interaction with the immune system and the inflammatory process. The same applies to all devices that are in contact with the inside of the body or those in contact with the skin, such as clothing, regardless of whether or not they are functionalized with fibers and synthetic or electronic components;

– body-machine integration: the body is a hierarchically organized structural and functional whole, highly integrated and stable (Chauvet 1992) in the space and time of its domain of stability or ecological niche. As an individualized "object", it is structurally discrete and functionally continuous in its relationship to the natural or artificial life environment on which it depends[20] (Nasa 2009). The machine, or

19 The environment or ecosystem can be described by several aspects: the abiotic aspect (non-living, geophysical and geoclimatic), the biotic aspect (relative to life and which is favorable to it and to all living organisms), the socio-cultural aspect (relative to intraspecific or inter-individual modes of interaction within a group) and the socio-technical aspect (relative to both individual or collective human organization, integrating artificial technological or deontic systems).

20 Which we summarize by "Who would have the idea of separating the goldfish from its water and its bowl?!"

technical individual, is an assembly of components resulting from a functional analysis, and a process of conception and assembly of discrete physical and logical elements that are more or less passive or active. Its coherence of form and structure, and its functional continuum, are ensured by the initial design or model, the choice of constituent materials and the method of production or assembly. It depends on the expertise of the designer(s) to think analytically and synthetically about the desired technical system for a given activity and environment. For human-machine systems, the system of interface and interaction with humans is also the subject of this engineering approach. The problem of body-machine integration is to artificially create a structural and functional continuum by assembling two objects that are distinct in nature and constitution but must be anatomically and physiologically compatible.

Currently, in technical fields such as aerospace or construction, the consideration of human beings in the human-system integration is reduced to biomechanical properties and categories of symbolic and computable cognitive behavioral variables called "human factors". Physiology is often only reduced to the physical and discrete measurement of variables such as heart rate, blood pressure, strength, elasticity, etc., and the statistical processing of data considered representative of the stochastic process being measured. Even if it is a safety or quality issue, the consideration of the human being is reduced to physical or cognitive ergonomics, and the "design" of physical or logical interfaces and interactions as tools or elements of the technical and managerial production process, and its more or less automated piloting or control systems.

Technology and the artificialization of our environment not only modify our behavior and our ability to perceive, decide and act, but they also condition the perception of our body, the conceptualization of its biological nature and our ability to think, heal or increase it.

The understanding of the physical-chemical, molecular, electro-magnetic or mechanical dimensions of living organisms has led to numerous advances in medical technologies for diagnosis and treatment. However, not everything is calculable and reducible to a discrete physical measurement. The physiological function is not reducible to a mathematical function. This was already understood by von Neumann before he died (von Neumann 1958). With computer science, starting from a physicalist reductionism, we have entered the area of cognitive and computational reductionism. In this framework, even if the promises and virtual developments of new technologies carry the hope[21] of an ideal world with infinite

21 Let us not forget that hope was the last of humanity's evils that was left in the box once Pandora opened it.

imaginable possibilities, reality resists[22]. Although we admit the power of mathematical abstraction in its capacity for generalization and intelligibility as necessary, it is not sufficient to fully apprehend and model the world, especially in the field of biology and medicine.

Logical-mathematical rationality without emotions is not viable (Damasio 2001). Art, as aesthetic expertise[23] and know-how, necessarily complements the formal modeling[24] of living systems and, in particular, models of a healthy or sick body. In the course of its history as a cultural and symbolic expression, art has shown us the evolution of the representation of body forms and the technology of exploration or treatment (Bordin *et al.* 2015). The relationships between art, medicine as an applied science and engineering are part of our cosmogony. They are fundamental to medical technology.

Understanding the concepts (Conant and Ashby 1970), the stakes of co-modeling "living object – artificial object" systems for conception and implementation for rehabilitation or treatment, for "life support" or maintenance in operational condition, such as in extreme situations for a patient in resuscitation or an astronaut during a long-term mission, helps us to think.

In these extreme conditions and environments, we can envisage the modes of existence of these technical objects (Simondon 1958) – the technical element, the technical individual and the networks of technical objects – in a process of technological and cultural co-evolution.

With Simondon, it is therefore an epistemological and technological challenge to go beyond the Aristotelian hylomorphism of form/matter dualism, where individualization is constituted by the encounter of form and inert matter, where existence precedes essence. For living forms in the space of species, there is no

22 "While everything sleeps, even the sick person, the disease is still awake," Cioran.

23 Aesthetics is the science of sensations, the foundation of emotions, and of what is a matter of knowledge (Fass 2002).

24 Formal methods are methods for the logical-mathematical specification of cyberphysical systems – programs or software and electronic machines (Fass 2015). They ensure the correctness of an abstract or formal model that adheres to a well-defined set of requirements, the specific needs or characteristics of the system to be modeled and the model of the technological artifact to be implemented or manufactured (Gibson and Méry 1998). Their language of representation is structured with clear and well-defined semantics, which can be used for the taxonomic verification of clinical indications and medical protocols to be executed automatically, as is the case for a pacemaker (Méry and Singh 2010), a drug injection pump or a neuroprosthesis.

way to predetermine phenotypes (Longo and Zakhama 2013). Phenotypes and ecosystems co-constitute and jointly produce the space of possibilities.

To understand the organization of living organisms, to imitate, model (simulate) or synthesize it, it is necessary to reposition the question of the form of living organisms in the continuum of phylogenesis, ontogenesis (morphogenesis) and epigenesis.

With Gilbert Chauvet, we think of form as being what represents the greatest organizational potential, in other words, structural and functional stability. There is an optimal stability that depends on the complexity of the system, the age and the living environment or ecosystem.

Imitating: "it's pretending …" Mimicking: "it's to act like ..." Synthesizing: "it's doing it for real ...", as the children say.

With the evolution of medical technology and the challenges of the human-system integration[25], a theoretical paradigm shift is necessary if we wish to invent or bring about the emergence of new artificial living forms for therapeutic purposes, or to allow the exploration of space by extending the domain of life.

11.2.3. *A conceptual and theoretical framework: the mathematical theory of integrative physiology (MTIP) by Gilbert Chauvet*

This theory, which first provides a conceptual framework, examines the hierarchical organization of structures (in other words, anatomy) and functions (in other words, physiology and behavior) of living systems. It allows the search for principles of optimality that ensure functional organization and its dynamics. MTIP introduces several principles:

– the functional hierarchy, based on structural organization in space scales;

– the functional organization in time scales;

– the structural units that are the anatomical elements in physical space, from the molecular to the organism.

To address the problem of structural discontinuity, MTIP introduces the concept of functional interaction ψ generated by a source-structure s to a sink-structure S as a coupling element between the physiological functions performed by these structures.

25 Which we have called (Fass 2002) "the individual-environment complex."

This scheme does not correspond to a classical interaction like the one we approach in physics. Here, the functional interactions present three fundamental properties at each level of organization:

– non-symmetrical, they lead to a representation by a non-oriented graph;

– non-local, they induce non-local fields;

– by coupling two structural hierarchical elements, they increase the functional stability of the living system.

Due to the constraints of the structural hierarchy, Gilbert Chauvet chose a representation that leads to specific biological concepts and stated the hypothesis that a biological system can be represented mathematically by a set of functional interactions of the $s \xrightarrow{\psi} S$ type. Here, the main question is to determine whether there is a cause for the existence of the functional interactions, in other words, the set of $s \xrightarrow{\psi} S$ triplets. And, if so, what could be the origin of the existence (identification) of s, S and ψ, which together form a $s \xrightarrow{\psi} S$ component of the system?

According to Chauvet, the principle of stabilizing self-association (PAAS) (Chauvet 1993a), which is the fundamental principle of MTIP, provides an answer: for each ($s\psi S$) triplet, noted as $s \xrightarrow{\psi} S$, where s is the source system, S is the sink system and ψ is the functional interaction, the stability domain of the $s \xrightarrow{\psi} S$ system is wider than the stability domains of s and S taken in isolation. In other words, increased complexity of the $s \xrightarrow{\psi} S$ system corresponds to increased stability. The MTIP consists of a representation: the set of functional $s \xrightarrow{\psi} S$ interactions, a general organizing principle – the PAAS – and a hypothesis – any biological system can be described as a set of functional interactions and structural units (sources, wells, non-locality, non-symmetry and field operator). The formal representation of a biological system has two sides: the functional organization of the formal biological system (O-FBS) (Chauvet 1993b) and the dynamics of the formal biological system (D-FSB) (Chauvet 1993c).

The functional organization of the formal biological system (FBS) involves the combinatorial properties of elementary functional interactions (sources, sinks and their interactions); their number corresponds to the degree of organization of the structure. This formal biological system can be represented by a hierarchical graph that models its topology.

Formal biological system dynamics (D-FBS) deal with the formal expression of the dynamic process, the evolution in time and space of the product emitted by a source, which acts remotely on a sink. The location of sources and sinks and the distance between them affect the time it takes for the signal to reach the sink. Thus, the geometry of the biological system, which this representation includes, influences the dynamics of the process: the spatial-temporal action of one structural unit on another.

Four types of constraints on physiological phenomena thus interact: the dual structural and functional hierarchy; the geometry of the system, in other words, the anatomy and histology that describe the organization and spatial distribution of structures, which induce delays in the transmission of functional interactions; the structural discontinuities that physically separate physiological processes; and the space-time continuum in which these processes evolve.

The formalization of this fundamental relationship, between a structural and functional organization, allows a three-dimensional representation of biological systems. The existence of general principles at the origin of this functional organization and of the uniqueness of the biological system as a hierarchical system are based on two hypotheses formulated by Chauvet: on the one hand, the self-association hypothesis, which determines the existence of functional interactions of the biological system and ensures the topological stability of their organization; and on the other hand, the optimal hypothesis, which defines the biological system as the form representing the greatest organizational potential. The optimal hypothesis is at the origin of the hierarchical organization of functional interactions, in which the functional order always increases, the system self-organizing in such a way that it increases its stability domain and functional order at the same time.

Gilbert Chauvet's MTIP provides a conceptual framework for researching the principles of optimization of functional organization and its dynamics. Applied to the conceptualization of the enhanced human, it provides a human-machine conception framework, allowing the search for the optimal structural and functional organization of biocompatible and biointegrable cyberphysical or cyberphysical-chemical artificial systems (Fass and Gechter 2015). This thus induces topological stability of the whole organized or extended system, increasing the stability domain and the functional order of the human-machine system in an integrated geometry.

11.2.4. *Forms of thinking or processing by machines*

"Can a machine think?" This question has been driving the scientific community since the 1950s. It has also tested René Descartes' animal-machine concept and Julien Offray de La Mettrie's "man-machine" concept. Before them, the myth of the Golem in Prague or, more recently, the story of Pinocchio written by Carlo Collodi, gave humanity to objects produced by technology. The functionalist and cognitivist conceptualization of thought, reducing it to the ability to produce reasoning or make a decision – provided that it is based on language and numerical or symbolic calculation – would lead us to consider whether or not machines imitating human behavior could be endowed with the ability to think. Humans have been lending magical properties to their artifacts or machines since the time of *sapiens*. It is an anxiolytic behavior with regard to our awareness of our finitude and infinity. Asking the question "What is it like to be a bat?", Thomas Nagel (1974) examines the philosophical question of the mind-body problem. Through this question, he proposes a questioning of the physicalist reductionism of consciousness and of the subject/object, individual/environment duality. To us, the issue of the body/thought duality seems to disappear with Simondon if, as we propose, we extend the theory of integrative physiology to the higher level of organization of the body to its natural or artificial living environment.

The fundamental problem for the conception of machines or medical technology, "life support" systems or systems to assist in maintaining operational condition and action in extreme environments, for example, is no longer about "Making a Mind versus Modeling the Brain" (Dreyfus and Deyfus 1988), but about how to imitate life in matter and in its ecosystem, and therefore how to think about the integrative, adaptive and evolutionary organization of life forms[26].

11.3. Simulation, biomimetics and bioprinting: a future for medical technology

Artificial morphogenesis and its modeling are a current research topic; they involve the design and production of new types of prostheses, as well as tissue bioengineering or art (Brayer and Zeitoun 2019). They concern artificial ontogenesis and epigenesis and human-artifact co-adaptation.

26 Thus, could we rephrase Nagel's question for integrative bioengineering: "What is being blind when you are clairvoyant in the dark?" and "What is being human when you are no longer subject to Earth's gravity?"

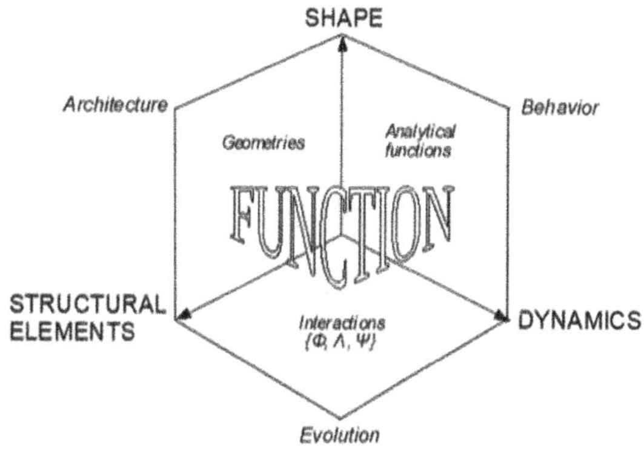

Figure 11.1. *Our isomorphic epistemic framework for design engineering and the integrative organization of human-artifact systems (Fass 2012)*

Additive digital manufacturing or 3D printing (André *et al.* 1984) is taking regenerative medicine, and the design of medical objects and devices, into a new era of modeling, simulation and systems engineering. Tissue engineering and bioprinting currently face two major challenges (Derakhshanfar *et al.* 2018; Leberfinger *et al.* 2019):

i) *biomanufacturing*, which involves techniques for the 3D fabrication of artificial biological objects by combining cells and biomaterials;

ii) in vivo integration of these substitutes, which implies biocompatibility and structural and functional (topological and geometrical) biointegration.

The modeling and digital simulation of artificial biological systems, in particular for tissue engineering and the production of custom living grafts or orthoses from biocompatible materials, is a pluridisciplinary scientific challenge for tissue engineering, and regenerative (National Academy of Sciences, Engineering and Medicine 2017) and predictive medicine. Currently, the theoretical models implemented by tissue engineering research are based on experimental knowledge or evidence in the tradition of Claude Bernard's experimental medicine, and focus on the impression of tissue from a single-cell type in two dimensions or in successive planes (Gu *et al.* 2016). While this work has led to very promising results, mainly in vitro (Henrionnet *et al.* 2017), it does not take into account the complexity of the hierarchical organization of a biological system or its evolution in vivo. This difficulty is of an epistemological nature: the concept of tissue, introduced by Bichat in his general anatomy, is interpreted by mechanistic thinking and, even worse, he reduces a biological tissue to an assembly of components (cells and matrix substrate), independently of its morphogenetic and epigenetic evolution that is relative to its functional interactions with these underlying (molecular) and overlying (organs and systems) levels of organization, in the relativity of scale and the totality of a given organism. The additive manufacture of tissue substitutes may reinforce this way of thinking, without critical analysis of these concepts.

It is therefore necessary to provide the means to direct the conception of artificial biological tissues through bioengineering, driven by revised concepts and evolutionary and "integrative" models. It is therefore necessary to go beyond the question of the essence of the "artificial biological tissue" object reduced to an assembly of its components, putting it at the service of system description through engineering analysis and modeling problems. We propose to base ourselves on the isomorphic epistemic framework of integrative design and organization (see Figure 11.1), in which function is an abstraction of the functioning that results or emerges from the construction and stability of the integration of structural elements, from form (design and also *Gestalt*, as a biologically meaningful form, for example, in tissue or immunological terms) and the dynamics of the created system. In a complementary way, it is necessary to make this notion practical through the establishment of a criteriology for tissue characterization and natural or artificial tissue bioengineering in biomimetic engineering (Fernandez-Yague *et al.* 2015; Dang *et al.* 2018). This set of biocompatibility and biointegration criteria will allow the description of matrix (scaffold) and cellular material complexes, through a design ontology of bioartificial tissues in their genesis and evolution in vitro and in vivo.

Thus, the challenge is to define a general framework for the formal modeling of any artificial biological tissue and a set of safe design criteria or design ontology. The challenge of bioartificial tissue modeling and production lies between theoretical and experimental integrative biology and formal computational methods. The safe design of bioartificial tissues and their realization by additive manufacturing must be based on epistemological reflection and on the modeling of a biocompatible and biointegrable artificial biological system simulating its morphogenesis and evolution. Indeed, for a 3D printer to generate a synthetic tissue, it is necessary to establish a digital pattern from a model. This multidimensional or multi-criteria pattern must be simultaneously based on the organization of three planes of specification: architecture, behavior and evolution. These planes of specification result from a combination of two-by-two of three axes of requirements: structural elements (cells, matrix biomaterials, growth factors, etc.), forms (Euclidean or non-Euclidean, geometry of the living, etc.) and dynamics (physical, logical and physiological interactions). These axes of requirements must themselves meet the specific needs of the conceptual and expert field of biology, integrative design and engineering of a natural or bioartificial anatomical-physiological system.

From the Vitruvian Man and his ideal proportions to the reality of the human condition of the body and living forms, medical technology will essentially be biomimetic and biointegrative.

11.4. References

André, J.-C., Le Mehaute, A., De Witte, O. (1984). Dispositif pour réaliser un modèle de pièce industrielle. Patent, no. FR8411241A.

Bailly, F., Longo, G. (2006). *Mathématiques et sciences de la nature. La singularité physique du vivant*. Hermann, Paris.

Bar-Cohen, Y. (ed.) (2006). Introduction to Biomimetics: The Wealth of inventions in Nature as an Inspiration for Human Innovation. In *Biomimetics, Biologically Inspired Technologies*. Taylor & Francis Group, Boca Raton.

Bernard, C. (1984). *Introduction à l'étude de la médecine expérimentale*. Flammarion, Paris.

Bordin, G., Bussagli, M., Polo d'Ambrosio, P. (2015). *Le livre d'or du corps humain, anatomie et symboles*. Hazan, Paris.

Brayer, M.A., Zeitoun, O. (2019). *La fabrique du vivant "Mutations/Créations 3"*. Centre Pompidou, Paris.

Cameron, D.E., Bascher, C.J., Collins, J.J. (2014). A brief history of synthetic biology. *Nature Reviews. Microbiology*, 12(5), 381–90.

Chauvet, G.A. (1992). Systèmes physiologiques. In *Systémique. Théorie et applications*, Le Gallou, F., Bouchon-Meunier, B. (eds). Lavoisier, Paris.

Chauvet, G.A. (1993a). Hierarchical functional organization of formal biological systems: a dynamical approach. i. The increase of complexity by self-association increases the domain of stability of a biological system. *Philosophical Transactions of the Royal Society of London B: Biological Sciences*, 339(1290), 425–444.

Chauvet, G.A. (1993b). Hierarchical functional organization of formal biological systems: a dynamical approach. ii. The concept of non-symmetry leads to a criterion of evolution deduced from an optimum principle of the (o-fbs) sub-system. *Philosophical Transactions of the Royal Society of London B: Biological Sciences*, 339(1290), 445–461.

Chauvet, G.A. (1993c). Hierarchical functional organization of formal biological systems: a dynamical approach. iii. The concept of non-locality leads to a field theory describing the dynamics at each level of organization of the (d-fbs) sub-system. *Philosophical Transactions of the Royal Society of London B: Biological Sciences*, 339(1290), 463–481.

Chauvet, G.A. (1995). *La vie dans la matière. Le rôle de l'espace en biologie*. Flammarion, Paris.

Clynes, M.E., Kline, N.S. (1960). Cyborgs and space. *Astronautics*, 14(9), 26–27, 74–75.

Colgate, E., Wannasuphoprasit, W., Peshkin, M. (1996). Cobots: robots for collaboration with human operators. *Proceedings of the ASME Dynamic Systems and Control Division*, 58, 433–439.

Conant, R.C., Ashby, W.R. (1970). Every good regulator of a system must be a model of that system. *International Journal of Systems Science*, 1(2), 89–97.

Damasio, A.R. (2001). Emotion and the human brain. *The Unity of Knowledge: The Convergence of Natural and Human Science, Annals of the New York Academy of Sciences*, 935(1), 101–106.

Damish, H. (1987). *L'origine de la perspective*. Flammarion, Paris.

Dang, M., Saunders, L., Niu, X., Fan, Y., Ma, P.X. (2018). Biomimetic delivery of signals for bone tissue engineering. *Bone Research*, 6, 25.

Derakhshanfar, S., Mbeleck, R., Xu, K., Zhang, X., Zhong, W., Xing, M. (2018). 3D bioprinting for biomedical devices and tissue engineering: A review of recent trends and advances. *Bioactive Materials*, 3(2), 144–156.

Destruhaut, F., Vigarios, E., Andrieu, B., Pomar, P. (2011). Regard anthropologique en Prothèse Maxillo-Faciale : entre science et conscience. *Chimères*, 75(1), 45–56.

Dreyfus, H.L., Dreyfus, S.E. (1988). Making a mind versus modeling the brain: AI at a crossroads. *Daedalus*, 117, 15–43.

Fass, D. (2002). Base de connaissances multimodales et esthétique de la connaissance, aide à l'action et réalité virtuelle : théorie et expérience. PhD thesis, Université Henri Poincaré – Nancy 1, Nancy.

Fass, D. (2012). Augmented Human Engineering: A Theoretical and Experimental Approach to Human Systems Integration. In *Systems Engineering – Practice and Theory*, Cogan, B. (ed.). IntechOpen, London.

Fass, D., Gechter, F. (2015). Towards a Theory for Bio-Cyber-Physical Systems Modelling. In *Digital Human Modeling. Applications in Health, Safety, Ergonomics and Risk Management, Lecture Notes in Computer Science*, Dufy, V.G. (ed.). Springer, Berlin, 9184, 245–255.

Fauchard, P. (1728). *Le chirurgien dentiste ou Traité des dents*. Jean Mariette, Paris.

Fernandez-Yaguea, M.A., Abbaha, S.A., McNamarab, L., Zeugolisa, D.I., Pandita, A., Biggs, M.J. (2015). Biomimetic approaches in bone tissue engineering: Integrating biological and physicomechanical strategies. *Advanced Drug Delivery Reviews*, 84, 1–29.

Finch, J. (2011). The ancient origins of prosthetic medicine [Online]. *The Lancet*, 377(9765), 548–549. Available at: https://doi.org/10.1016/S0140-6736(11)60190-6.

Gibson, J.-P., Méry, D. (1998). Teaching Formal Methods: Lessons to Learn [Online]. In *Irish Workshop For Formal Methods 1998*. Cork, Ireland, available at: https://hal.inria.fr/inria-00098616.

Gombrich, E.H. (1996). *L'art et l'illusion*. Gallimard, Paris.

Gu, B., Choi, D., Park S., Kim, M., Kang, C., Kim C.H. (2016). 3-dimensional bioprinting for tissue engineering applications. *Biomaterials Research*, 20, 12.

Guyatt, G. *et al.* (1992). Evidence-based medicine, a new approach to teaching the practice of medicine. *JAMA*, 268(17), 2420–2425.

Heard, E. (2013). Épigénétique et mémoire cellulaire [Online]. In *Annuaire du Collège de France 2012-2013. Résumé des cours et travaux 113e année*. Available at: https://journals.openedition.org/annuaire-cdf/2420.

Henrionnet, C., Liang, G., Roeder, E., Dossot, M., Wang, H., Magdalou, J., Gillet, P., Pinzano, A. (2017). Hypoxia for MSC expansion and differentiation: the best way for enhancing TGFß-induced chondrogenesis and preventing calcifications in alginate beads. *Tissue Engineering*, Part A, 23(17/18), 913–922.

Hwang, J., Jeong, Y., Park, J.M., Lee, K.H., Hong, J.W., Choi, J. (2015). Biomimetics: forecasting the future of science, engineering, and medicine. *International Journal of Nanomedicine*, 10, 5701–5713.

Leberfinger, A., Dinda, S., Wu, Y., Koduru, S., Ozbolat, V., Ravnic, D., Ozbolat I. (2019). Bioprinting functional tissues. *Acta Biomateralia*, 1(95), 32–49.

Lechermeier, G. (2015). Définition du vivant et émergence de la vie : entre rupture et continuité, saisir l'originalité du vivant. PhD thesis, Université Paris 1 Panthéon-Sorbonne, Paris.

Longo, G. (2010). Interfaces de l'incomplétude [Online]. *La Matematica*, 4, 219–262. Available at: https://www.di.ens.fr/users/longo/Papers/Incompletude.pdf.

Longo, G., Zakhama, N. (2013). Le modèle comme regard organisateur du réel. In *Exposé invité, Colloque "Naturaliser le modèle"*. Chambord, 14 October.

Lorenz, K. (1984). *Les fondements de l'éthologie*. Flammarion, Paris.

McCarthy, J., Minsky, M.L., Rochester, L., Shannon, C.E. (1955). A proposal for the Dartmouth summer research project on Artificial Intelligence [Online]. Available at: http://www-formal.stanford.edu/jmc/history/dartmouth/dartmouth.html [Accessed October 2019].

Méry, D., Singh, N.K. (2011). Technical Report on Interpretation of the Electrocardiogram (ECG) Signal using Formal Methods [Online]. Technical report, Nancy. Available at: https://hal.inria.fr/inria-00584177/file/TechRepoECG2011.pdf.

Nagel, T. (1974). What is like to be a bat. *The Philosophical Review*, 83(4), 435–450.

NASA (2009). Human Health and Performance Risks of Space Exploration Missions. In *Evidence Reviewed by the NASA Human Research Program*, McPhee, J.C., Charles, J.B. (eds). National Aeronautics and Space Administration, Lyndon B. Johnson Space Center, Houston.

National Academies Keck Futures Initiative (NAKFI) (2010). *Synthetic Biology: Building on Nature's Inspiration: Interdisciplinary Research Team Summaries* [Online]. The National Academies Press, Washington. Available at: https://doi.org/10.17226/12836.

National Academy of Sciences, Engineering, and Medicine (2017). *Navigating the Manufacturing Process and Ensuring the Quality of Regenerative Medicine Therapies* [Online]. The National Academies Press, Washington. Available at: https://doi.org/10.17226/24913.

Paré, A. (1585). *Les œuvres d'Ambroise Paré* [Online]. Gabriel Buon, Paris. Available at: https://www.nlm.nih.gov/exhibition/historicalanatomies/pare_home.html.

Prochiantz, A. (2012). *Qu'est-ce que le vivant ?* Le Seuil, Paris.

Roco, M.C., Bainbridge, W.S. (2003). *Converging technologies for improving human performance: Nanotechnology, Biotechnology, Information Technology, and Cognitive Science* [Online]. National Science Foundation, Kluwer Academic Publishers, Dordrecht. Available at: www.wtec.org/ConvergingTechnologies/Report/NBIC_report.pdf.

Rosenblueth, A., Wiener, N. (1945). The role of models in science. *Phylosophie of Science*, 12(4), 316–321.

Schmitt, O.H. (1969). Some interesting and useful biomimetic transform. In *Proceedings of Third International Biophysics Congress of the International Union for Pure and Applied Biophysics*. Boston.

Simondon, G. (2012). *Du mode d'existence des objets techniques*. Aubier-Montaigne/ Flammarion, Paris.

Turing, A. (1950). Computing machinery and intelligence. *Mind*, 50, 433–460.

Vincent, J.F.V. (2001). Stealing Ideas from Nature. In *Deployable Structures*, Pellegrino, S. (ed.). Springer, Berlin, 51–58.

Von Neumann, J. (1958). *The Computer and the Brain*. Yale University Press, New Haven.

Wiener, N. (1948). *Cybernetics, or Control and Communication in the Animal and the Machine*. MIT Press, Boston.

12

From Living to Thinking: Mosaic Architecture

Georges CHAPOUTHIER
ICM-CNRS, Paris, France

12.1. Introduction

How is the complexity of living beings constituted? Do living forms present analogies that might lead one to think that great universal principles govern their construction? Alongside (and in addition to) the omnipresent Darwinian principles of natural or sexual selection, which of them lead to the evolution of living things? Are there any great general principles, inherent to living matter, which condition the biological evolution towards a certain complexity? This is what this chapter would like to demonstrate. We will defend the idea that two great principles, inherent to the living world, also lead, in addition to Darwinian principles, to the complexity of living beings[1]. These two principles lead to that which is called a "mosaic complexity".

12.2. Two main principles

Indeed, two main principles seem to be involved in such a process. We will call them the "principle of juxtaposition" and the "principle of integration". They are presented in Figure 12.1.

---

1 In this chapter, we have favored the use of the term *complexity*, which we believe is more explicit than the often-used concept of "emergence". However, it is clear that the two concepts do overlap.

The Explosion of Life Forms,
coordinated by Georges CHAPOUTHIER and Marie-Christine MAUREL. © ISTE Ltd 2020.

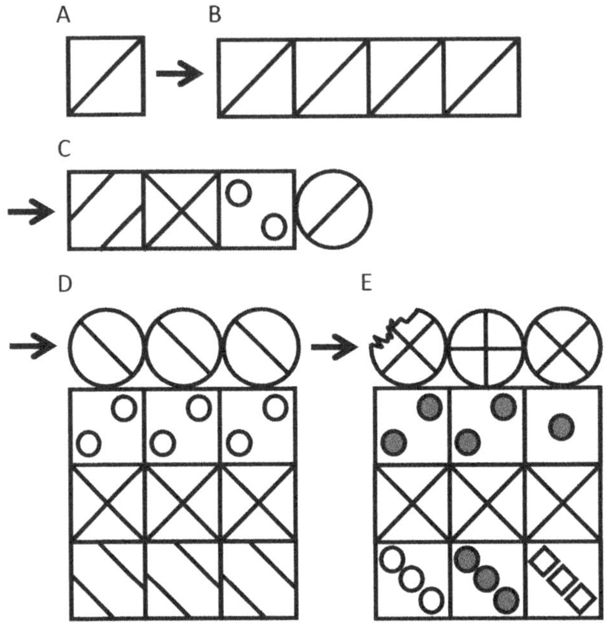

Figure 12.1. *Schematic presentation of the mosaic complexity model. Units (A) are first juxtaposed to form the structure (B), where all units are similar. The units are then integrated to form the structure (C), of which the original units then become part of. The process can be repeated by juxtaposing structures (C) to form (D), and then integrating them to form (E), which then becomes a "mosaic of mosaics" (source: Chapouthier 2001)*

In theory, entities could be combined indefinitely by juxtaposing them and then integrating them to form stages of increasing complexity. In practice, in the living world as we see it, between the organelles that make up cells and animal populations, we can, in fact, distinguish five or six stages of increasing complexity.

12.3. Genes and cells

Let us first show how the anatomy of living beings is constructed from these two main principles (Chapouthier 2001, 2018).

Firstly, with regard to genes, we know that in silent intron zones, occasional duplications of identical introns occur and that these duplications can be repeated (which, according to our model, can appear as a juxtaposition of identical introns). Secondly, some of these introns can be modified by mutations. The set of these mutations thus results in groups of modified introns (which, according to our model, appears as an integration). When such a group of introns is sufficiently integrated, it can then become functional in the form of exons and produce a new attribute of the organism under consideration. If this attribute proves useful for the survival of the organism, it will be preserved by Darwinian selection. If not, it will disappear and this exon test will be discontinued. Authors such as Susumu Ohno (1970) considered that such processes of silent modification of introns, followed by their subsequent emergence as exons, constituted the very mechanism that allowed the evolution of species.

Let's move on to cells. Cells can be juxtaposed, all alike, to form tissues or independent sets like *gonium*, where all cells retain identical properties. In the *Volvox* algae, on the other hand, there are different types of cells: ordinary cells and reproductive cells, which indicates the beginning of integration. Of course, this integration between different cell types, initiated here in the *Volvox*, will be further enhanced in most multicellular plants and animals (metaphytes and metazoa). It should be noted that it is specifically in *Volvox* that the Richard E. Michod group (Michod 1999, 2009), at the University of Arizona (US), has carried out research on "individuality mutations", in other words, on the causes, particularly biochemical causes, which move from a level of complexity, such as a cellular one, to a more integrated level, which is multicellular here.

Still in this same field of transition from a unicellular to a multicellular level, interesting studies have been carried out by Paul B. Rainey's group and colleagues (Libby and Rainey 2013). These authors define three possible paths from a unicellular entity to a multicellular structure. The first path involves cell aggregates that remain stable and reproduce as aggregates of juxtaposed cells, by a kind of asexual cutting. The second path uses aggregates that regularly reconstitute themselves from isolated cells. An example is the (sexual) reproduction of metazoans, based on the soma/germen distinction. The third path involves successive phases, sometimes consisting of isolated cells and sometimes of aggregates. The authors suggest that during the evolution of species, paths 2 and 3, which allow a gradual appearance of multicellular forms, are more plausible.

The same team (Hammerschmidt *et al.* 2014, pp. 75–79) attempted to investigate these possibilities experimentally using the *pseudomonas fluorescens* bacterium, which sometimes develops multicellular structures. A mutation causes isolated cells to overproduce an intercellular glue, leading to layers of juxtaposed cells. This multicellular structure obtains better access to oxygen. However, it has a short lifespan, as "cheating cells" soon appear, which take advantage of the biochemical advantage of the cell layer, but do not participate in its integrity and do not maintain intercellular adhesion. According to the authors, processes that suppress cheating cells are probably at work in the multicellular structures that are expected to maintain themselves. The authors provided an experimental demonstration of this: they were able to subject their preparation of *pseudomonas fluorescens* to a cheating-cell suppression regime and thus obtained an increase in the lifespan of the multicellular layers.

12.4. More complex anatomical mosaics

When we now look at two-layered animals, we see that polyps that are all identical to each other can juxtapose themselves over thousands of kilometers to form large coral reefs. Colonies of integrated polyps, known as siphonophores, are also known to float in the oceans. In these colonies, the polyps have specialized to become floating, nurturing, defensive or reproductive. We thus pass from juxtaposed forms to integrated forms.

As far as three-layered animals are concerned, the earthworm in our gardens is a fairly good example of a juxtaposed structure. The metamers, of which it is composed, are (with the exception of a few details) identical to each other. If we now integrate this juxtaposed structure, we get the bee, the octopus, the goldfish or the chimpanzee, which are integrated three-layered structures. The anatomical integration is not always perfect. Thus, the abdomen of insects reveals, by its metamorphosis, its "juxtaposed" origin. Similarly, in vertebrates, the segmentation of the ribs and vertebrae reflects the same juxtaposition within an integrated organism.

Can we, beyond the three-layered animals, envisage a new, more complex stage of which three-layered beings would become part? Can we imagine a juxtaposition and then a new integration of organisms that are already integrated, such as vertebrates? Juxtaposition exists. It is found in "Siamese twins", which constitute an anatomical juxtaposition of two identical three-layered individuals. However, the process does not seem to have gone any further, perhaps because the mobility of these entities makes integration difficult. At this level, as we shall see below, it is necessary to move on to phenomena of *social* juxtaposition and integration.

12.5. Epistemological rehabilitation of asexual reproduction

It should be noted in passing that this insistence on mosaic complexity epistemologically rehabilitates the usefulness of asexual reproduction.

Darwinian selection made it possible to emphasize the immense benefit to the evolution of species of sexual reproduction, which can be effectively addressed by selection processes. The present model of mosaic complexity, on the other hand, allows the rehabilitation of asexual reproduction. Complexification processes, which involve juxtaposition and integration, are first of all based on the stage of juxtaposition of entities, produced by asexual reproduction, and the like. This stage is in fact the result of the *non-separation* of entities that could have separated in the first place. The multiplication of cells of a tissue, the cutting of polyps of a coral reef, as well as metamerism or Siamese twins, are strictly asexual reproduction phenomena. The model of mosaic complexity thus makes it possible to underline the considerable importance for the evolution of species, and its march towards a certain complexity, of these asexual phenomena, which epistemology tended to neglect.

The model of mosaic complexity is perfectly compatible with Darwinian selection, which remains, of course, thanks to the diversity that sexual reproduction allows, an essential cause of the evolution of species. However, the mosaic complexity model is not directly based on Darwinian hypotheses, since it assumes the existence of major principles for producing complexity, which are internal to the functioning of living organisms. It can be considered compatible but complementary to the Darwinian hypotheses.

These conceptions can be compared to those of the Brazilian philosopher Paulo C. Abrantes (2011, pp. 395–408). Abrantes analyzes what he calls "transitions in individuality", in other words, transitions from a simple individual level to a larger level that integrates individual entities into a larger entity, which fits perfectly with our mosaic model through juxtaposition and integration. Abrantes shows that, in order to make this transition from a low level to a higher level, it is necessary to go through processes he calls "de-Darwinization", in which the lowest level is no longer governed by Darwinian rules. In other words, without excluding the general action of the Darwinian hypotheses in the evolution of the living, Abrantes supposes that, in the construction of complexity, the transitions in individuality are moments when the parts of a whole are no longer strictly subjected to the mechanics of selection, and only act again when the parts of the whole have been integrated into a higher-ordered entity.

12.6. Social mosaics

Let's go back to our two basic principles and show how, in animals, they can also be applied at a social level.

In various animals, we find phenomena known as "crowding", where individuals group together while maintaining identical behavioral patterns. Thus, when birds group together to sleep, all individuals can be considered to be of a similar social level. It is therefore a social juxtaposition of individuals. However, in the case of colonies or societies, there are phenomena of integration. In termite, ant or bee colonies, as well as in carnivore or primate societies, individuals have very different behaviors and social functions, often linked to a sharing of tasks and a hierarchy. Of course, the same is true for the societies of a particular primate, the human being.

In human societies, social integration can have different degrees. The degree of autonomy of individuals (who are the "tesserae" of the social mosaic) in relation to society (the "whole") constitutes what is usually called "freedom", which is greater in so-called democratic societies than in totalitarian societies.

12.7. Encephalic mosaics

Mosaic complexity is not only evident in the general anatomy or socialization of animals. It also affects the particular anatomy of certain organs. We would like to give three examples of this in the case of vertebrate brains. We will not go into the example, which is striking but still little known, involving the juxtaposition and then integration of nerve ganglia in cephalopod mollusks.

Firstly, with regard to the vertebrate brain as a whole (see Figure 12.2), it is known that during embryogenesis, in the anterior part of the nerve tube, it is formed by five vesicles juxtaposed from front to back. In the most anatomically primitive vertebrates, such as some fish, the five vesicles can easily be found, slightly modified, and aligned. In the more anatomically evolved vertebrates, this juxtaposed structure is modified by the dilation of the elements of the first vesicle, the telencephalon (specifically the cortex and the striatum), or of the fourth vesicle (the cerebellum). The result, particularly in mammals and birds, is compact and integrated organs, where it is not always easy to recover the original structure of five juxtaposed vesicles. In mammals, for example, the enlarged cerebral cortex of the first vesicle is divided into two right and left hemispheres, which include the elements of the base of the telencephalon (striatum) and even the elements that make up the second (diencephalon) and third (mesencephalon) brain vesicles. There is a clear shift from a juxtaposed structure to a highly integrated structure.

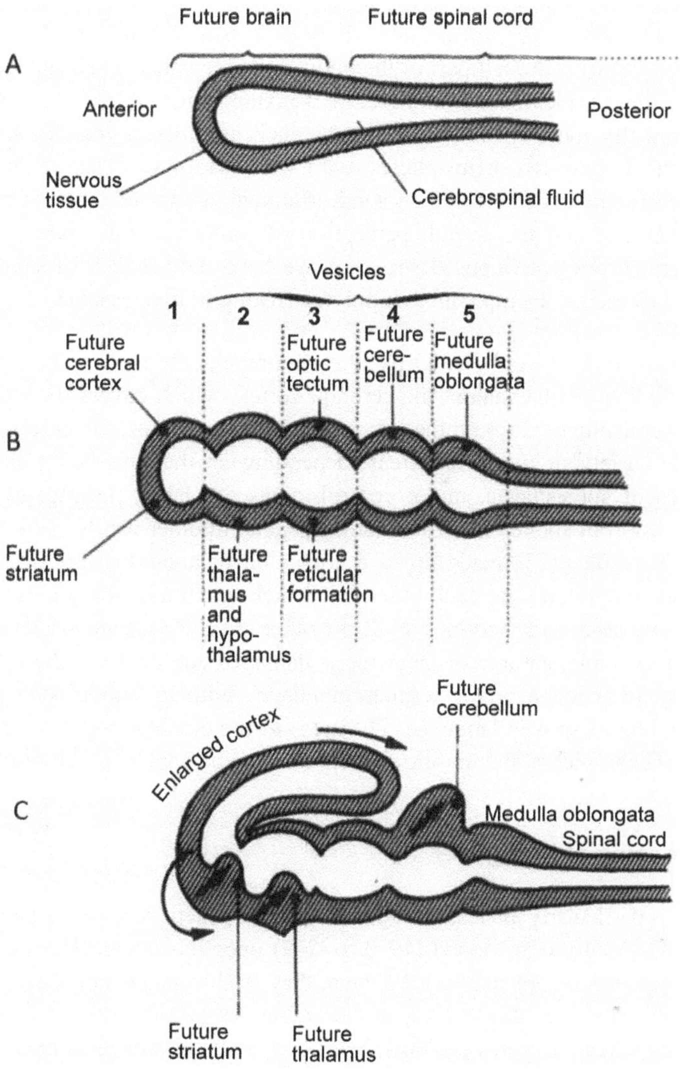

Figure 12.2. *Development of the brain in the vertebrate embryo. The vesicle (A) located in the anterior part of the nerve tube (shown here as a cross-section) transforms into five juxtaposed vesicles (B) and finally gives rise, by integration (C), to the structures of the final brain (source: Chapouthier and Matras 1982)*

A second example can be given by the two cerebral hemispheres. In the most anatomically primitive vertebrates, they are simply juxtaposed. The right hemisphere controls the left part of the body, and vice versa. Their functions are, for each hemisphere, identical. It is therefore a juxtaposition reduced to two elements. In anatomically more evolved vertebrates, such as birds or mammals, we see an integration of these two hemispheres, with specialization of one or the other in a particular function, such as a bird's song or human language, generally being in the left hemisphere, and the overall perception of forms generally being in the right hemisphere. From a juxtaposed structure, we have moved to a structure with two integrated elements, a simple mosaic only consisting of two tesserae.

Lastly, as a third example, let us take a more specific look at the dorsal part of the mammalian telencephalon, the cerebral cortex, which covers the surface of the cerebral hemispheres. In both the left and right hemispheres, the cortex is made up of a wide variety of functional areas, depending on the case being on the left or right: area of somesthesia, motor control, areas of vision, hearing, expression or comprehension of spoken or written language, etc. In other words, the surface of the cerebral hemispheres is made up of a collection of areas that are juxtaposed and sometimes integrated with each other, but which retain autonomy in relation to the cortex as a whole, and therefore to all the other areas. Evidence of this can be seen when, for example, an area of language is damaged and deprives the subject of the possibility of understanding written language, without suppressing his or her understanding of spoken language. The areas of the cerebral cortex are perhaps one of the best examples of the mosaic concept that can be applied to a living organism.

12.8. Mosaics of thought

What is particularly interesting is that the mosaic pattern can also be found in mental phenomena. As Jacques Monod (1970, p. 279) once remarked: "Ideas have retained certain properties of organisms. Like them, they tend to perpetuate their structure and multiply it. Like them, they can merge, recombine, segregate their contents, as well as evolve, and selection, in this evolution, without a doubt, plays a great role." In the same way, mental processes obey the same principles of complexity that we have analyzed in biological processes. We will first show this in the three examples of memory, language and consciousness before extending the subject to other consequences of human or animal psychology: the making of objects and cultural traits.

It can be shown that memory is not a single category, but a collection of very different abilities that have emerged as animal species evolved (Chapouthier 2006). Various memory tests can be done on animals from the main groups. These may be very simple memory tests (such as habituation or the tendency to alternate), more complex memory tests (such as Pavlovian or Skinnerian conditioning) or even more elaborate memory tests (such as semantic memories of spatial choices or cognitive rules). We can then discover that the most frustrating memories are found in most animal groups, whereas conditionings only appear in three-layered animals. An important detail: the more elaborate conditioning memories do not replace the simpler memories, but complement them. In fact, we can witness a juxtaposition of the different aptitudes.

The same is true if we look to semantic memories. They are only found in animals that are strongly "cephalized": vertebrates and cephalopod mollusks. Here again, they do not replace other types of memory, but complement them. An octopus or a crow, capable of spatial or cognitive memories, is still equipped with habituation, alternation or conditioning memories. All these abilities are juxtaposed and (rather weakly) integrated. The same is true in the human species, where a task such as swimming is learned through conditioning, and where mathematical problem solving involves semantic memories.

Let's look at language (Robert and Chapouthier 2013, pp. 211–223; Robert 2018). According to the theory of enunciation, when I pronounce a sentence, I juxtapose semantic units (words) in time, whose meaning (integration) only gradually appears gradually as the sentence progresses. The definitive meaning (the semantic integration of the different words) is only obtained when I have finished enunciating my sentence. This idea of a mosaic in language was taken up by Hombert and Lenclud in their summary on the origins of language in humans (Hombert and Lenclud 2014).

The integration that takes place during the sentence can, however, lead to reversals, a possibility that humor or literature can greatly take advantage of. For example, if I say: "this delicious kiwi... is covered with feathers", I begin by integrating the meaning of a fruit and end with that of a bird. There are countless examples in literature and poetry (Chapouthier 2013, 2018a). We will choose a haiku, which is a traditional Japanese poem consisting of three verses of five, seven and five syllables. Here is a haiku written in French by the poet Jean Monod (Antonini 2003, p. 45):

L'absente de tout
bouquet la voilà me dit
en se montrant l'aube

(Absent of any
bouquet she is here to tell me
dawn as she appears)

The three verses of the haiku will be considered here as the tesserae of the poem, although these could, in turn, be broken down into simpler semantic elements. In French, the first verse uses feminine words to immediately embody a woman who is "absent from everything". The second verse, which refers to a bouquet, makes one think of a flower. However, it is the third verse that gives meaning to the whole: it is in fact the dawn. Of course, the evoked images of the woman and the flower retain their relevance in the final integration and give the haiku its poetic aspect.

More generally, in language, the very ambiguity of words and their polysemy explain why an assertion or concept often needs to be clarified by additional sentences. To enable conversation, the mosaic of language must be constantly refined by giving it new tesserae.

There is also an appeal to the mosaic structure, in another form, as shown in linguist William S.Y. Wang's work (2007, pp. 8–16). In a study that explicitly refers to the "mosaic of language and its biological basis", Wang describes language as a multi-layered mosaic, where cultural layers are superimposed on biological layers over the course of life. This is therefore not a semantic analysis like the one Stéphane Robert has done, but Wang's point of view is clearly complementary to it.

Three other human skills can be related to language: music writing, drawing and mathematics. Let's give some examples. As far as music writing is concerned, Australian musicologist Marshall Heiser of the Queensland Conservatorium of Music, Griffith University (Heiser 2012), conducted a study of Brian Wilson's aborted album *SMiLE*. Heiser was able to conclude that the recordings had a mosaic-like structure where "the forms reveal characteristics that are not inherent to their parts". Heiser explicitly cites our mosaic model and its processes of juxtaposition and integration, ultimately calling this musical work "Brian Wilson's musical mosaic". As for drawing, if we follow René Baldy's (2011) explanation of the emergence of complex drawings in children, we find that children begin by juxtaposing "lines" or "circles" and then integrate these two forms to form more complex forms, such as a sun or a face, of which the lines and circles become part.

Mathematics (or more specifically, mathematical language) is a form of language spoken and written in symbols, which makes it possible to mimic the elements of the world and make them intelligible. Again, it is based on a juxtaposition of algebraic symbols or geometric propositions, which lead to integrations that constitute the result or conclusion.

Lastly, consciousness, which initially appears to us as something particularly unitary, to the point of constituting the very unity of our personality, is in fact a mosaic of modules juxtaposed and then integrated. We have a first experience of it at night, when we dream. The warped consciousness of the dream calls upon different modules from those we use during the day. But an even more striking example is so-called “split brain” subjects. These subjects have had, by accident, a rupture of the corpus callosum, the bundle of nerve fibers which, in a normal subject, ensures the transmission of information between the right and left hemispheres. Thus, a person with a split brain (supposedly right-handed) will say (with his left hemisphere, that of language): “I want to take my red coat with my right hand, but I can’t help it, my left hand takes my blue coat.” The (language-related) command of the left cortex gives a command to the right hand, while the (silent) command of the right cortex gives an opposite command to the left hand. Such antagonisms between right and left can also occur in normal human subjects, but they are immediately corrected by information passing through the corpus callosum. In subjects with split brains, they can no longer be corrected. These subjects have two separate personalities and two separate consciousnesses within them. The inter-hemispheric integration of the consciousness of the normal subject has given way to a pathological return to two juxtaposed faculties.

12.9. Man-made objects

Ludwig von Bertalanffy had proposed (von Bertalanffy 1968) a “general theory of systems”, according to which parallel or similar systems should exist in different fields of knowledge. If this is the case, it should be expected that the mosaic model of complexity could find applications in scientific fields other than biology or its psychological consequences.

In these extensions, however, it is necessary to contrast areas that have something to do with human psychology with those that are totally different from it.

In the former, one cannot be surprised that similarities or isomorphisms with the living can exist, since they are based, in a way, on human thought, of which the architectural kinship with living beings has been seen above. This is, for example, the case of man-made objects and, more recently, robotics, that is to say, the often-animated beings that the human species is now making and which, in many ways, resemble living beings.

As far as the objects are concerned, they are generally built on the basis of a first juxtaposition (for example, to make a watering can, the reservoir, the handle and the spout are juxtaposed), followed by an integration (the form evolves to integrate the three components; the handle may be part of the reservoir and the whole will adopt a harmonious and elliptical shape, better adapted to the expected function). Frédéric Kaplan (2013, pp. 17–18) rightly writes that “like the evolution of organisms, technical evolution proceeds through two complementary processes of juxtaposition and integration. A new whole is first created by combining elementary technical objects". This juxtaposed whole is, at first, weakly integrated, but then technical maturation produces "a coherent and unified whole". The same is true of much more complex objects such as machines and computers. Pierre-Yves Oudeyer (2015, pp. 206–207) notes that it was "in the early 2000s that a new technical step was taken, constituting a new stage in the organization of machines into systems: the juxtaposition and large-scale integration of these dynamic behavioral objects", which even allowed populations of robots to interact, and develop cultural and linguistic conventions among themselves, similar to those found in human beings.

Other human technical creations are made up of architecture and urbanism. The architect Denis Laming writes on this subject (Laming 2015, p. 59): "Cities [...] are markers of civilization: they are born, develop, become more complex and finally die. This ontogeny, followed by entropy, is a process that has a lot in common with the Universe, the living and artificial intelligence, and follows the model of the mosaic imagined by Georges Chapouthier." As we can see, many human creations can be analyzed according to the mosaic model of complexity.

12.10. Human and animal cultural traits

In more general terms, cultural traits, in other words, the behavioral traits that are transmitted between individuals through imitation or education, without passing through genetic transmission, are mosaic traits. Examples can be found in both human and animal cultures (Lestel 2001). These traits include the creation of tools (which, in animals, is similar to the particular case of the man-made objects mentioned above), but they also include animal or human communications, often non-language based, of moral rules within a group or aesthetic preferences for forms, colors or song patterns (Chapouthier 2009). All of these behavioral traits can be broken down into sequences of gestures juxtaposed in time, which lead to the realization of the trait through the integration of all these gestures (Chapouthier 2018b).

Let us look at the particular case of human morality. While animal morals only manifest themselves through gestures, such as helping the handicapped or reconciliatory behavior (De Waal 1989), human morals include discursive precepts that bring us back to language. But, in a much more precise way, the philosopher Vanessa Nurock (2018) has developed an analysis of the mainsprings of human morality, in which she appeals to notions of mosaic complexity in two areas. First, she shows a plurality of moral conceptions which, without excluding a "universal moral competency", leads to diversity, the juxtaposition of the morals we know. The second area analyzed by Nurock is that of our moral abilities. At the base of moral sense is empathy, which Nurock proposes to subdivide into three tesserae: "agentive empathy" (I have empathy for the other) and "emotional empathy" (I share the other's emotions), situated at a naive level, and "situational empathy" (I perceive the details of the other's situation), situated at a more reflective level. The interest of this model is to show that certain psychiatric disorders (autism, psychopathy, etc.) result in the weakening of some of these tesserae by sparing the others. The existence of the naïve-level empathic tesserae also makes it possible to account for childish morals and animal morals, which are less structured in terms of reflection.

12.11. A universality of mosaics?

Can we extend the concept of mosaic complexity to completely different areas of the biological horizon, and all its psychological or cultural consequences that we have just outlined? Can we imagine that this mosaic complexity, formulated on the basis of the architecture of living beings, can also be applied to the architecture of the universe in general? The philosopher Denis Forest had rightly questioned the model of the complexity in mosaic by asking if "such a form of organization is universal, [or if] it is often found (among others) in the most varied fields" (Forest 2018, p. 58).

Even if it is always difficult to aim for universality, one can take the risk if one adopts the Aristotelian conception, developed today by the neo-Aristotelian school of Konstantin Khroutski in Russia[2], according to which the universe is built according to a logic of the living. To say this in modern terms is obviously not to say that the universe is a great ape like King Kong, but to think that the architecture of complex structures everywhere in the universe is the same as that which we find on Earth, within the most complex systems that we are given to observe: living beings. Therefore, according to this Aristotelian thesis, there is a construction isomorphism between the macrocosm (cosmic) and the microcosm (terrestrial).

In fact, this thesis is supported by the reflections of the astrophysicist Jean Audouze. The latter rightly noticed (Audouze 2015, p. 45) the mosaic-like structuring of the cosmos. "Let's imagine," he tells us, "that the Universe is a huge Russian doll" made up of everything from super clusters of galaxies to objects in our terrestrial environment, including galaxies, stars and planets. "Each of these structures is made up of all the lower-ranked structures, and they only share part of their properties, although it has been shown that the behavior of the largest structures in the Universe reflects particle physics in some way. All these structures,

2 Available at: http://en.biocosmology.ru/electronic-journal-biocosmology---neo-aristotelism).

finally, fall under the 'mosaic model'". Without a doubt, we can therefore risk, with all the caution that this imposes, proposing a general Aristotelian thesis of mosaic complexity, outside the living world.

12.12. Conclusion: philosophical foundations

The set of examples given in this chapter shows, in a wide variety of fields of knowledge, the relevance of the mosaic model of complexity. In biology, this model offers an interpretation of the complexity of living beings, based on asexual reproduction and complementary to Darwinian selection, by proposing an internal determinism of the evolution towards complexity. It is also by their very construction that living beings become more complex. The model also adapts to the psychological and cultural properties of living beings, both human beings and (other) animals. The philosopher Arthur Saniotis (2013, pp. 334–340) believes that the mosaic model offers an integral approach to understanding biological processes. Lastly, it seems that the model can be generalized to the complex systems that populate the cosmos.

As for the philosophical positions that may underlie our theses, we have already noted two: von Bertalanffy's general theory of systems and Aristotle's theses on the relations between macrocosm and microcosm. These two philosophical positions, each in its own way, allow us to understand why properties, such as mosaic construction, are found in different disciplinary fields. We would like to add a third position here: that defended by Herbert Spencer.

The philosopher Jean-Claude Dupont (2018, p. 45) is right to bring our theses closer to those of Spencer. For Spencer, Dupont reminds us: "Integration means that, under the influence of the environment, the higher living forms group together elements that were found in the individual state in the lower forms." And, according to Spencer, these processes, which affect the evolution of living beings, also apply to the mind and matter. This is indeed very close to the theses that have been developed in this chapter, even if Spencer did not analyze the different stages of living beings, the brain and memory in as much detail as we do, or the precise architecture of language and morality as Stéphane Robert and Vanessa Nurock did. Conversely, the debate on the mosaicism of living things can indeed be brought as an argument, being rich in examples drawn from modern biological knowledge, and in favor of the Spencerian system itself.

The comparison with Herbert Spencer makes it possible to emphasize an essential point: mosaics are plastic and susceptible to evolution. Living organisms, thoughts, languages, cultural traits, objects, etc.: all of the above are necessarily part of evolutionary processes. This remark is a response to a pertinent criticism of mosaics by Jean-Hugues Barthélémy, who regretted the lack of insistence on their evolution (Barthélémy 2018, p. 35): "Any structural thought," Barthélémy told us, "even if it is intended to be cosmogenetic, is insufficiently genetic – in the reproductive sense". The reminder of the Spencerian kinship of our theses allows us to underline the plastic and moving, and therefore reproductive, aspect of the mosaics, for which I had even mentioned the term "dialectic" in another study (Chapouthier 2018a, pp. 40–41). This term is appropriate here, both in its materialistic sense of the "dialectic of nature" and in its Hegelian sense of the "dialectic of thought".

As we have shown, mosaics characterize both the architecture of the materiality of living beings and that of the thought they manifest.

12.13. References

Abrantes, P.C. (2011). Culture and transitions in individuality [Online]. In *De Araujo Dutra*, Meyer Luz, L.H. (ed.). Available at: https://issuu.com/ filosofia.ufsc/docs/rumos11.

Antonini, J. (2003). *Anthologie du haïku en France*. Aléas, Lyon.

Audouze, J. (2015). L'univers. In *Mondes mosaïques (Des astres à la ville, du vivant aux robots*), Audouze, J., Chapouthier, G., Laming, D., Oudeyer, P.Y. (eds). CNRS, Paris.

Baldy, R. (2011). *Fais-moi un beau dessin*. In Press, Paris.

Barthelemy, J.H. (2018). De l'ontologie au droit : quels fondements ? *Revue philosophique de la France et de l'étranger*, 1, 31–41.

Chapouthier, G. (2001). *L'homme, ce singe en mosaïque*. Odile Jacob, Paris.

Chapouthier, G. (2006). *Biologie de la mémoire*. Odile Jacob, Paris.

Chapouthier, G. (2009). *Kant et le chimpanzé, essai sur l'être humain la morale et l'art*. Belin, Paris.

Chapouthier, G. (2013). Des parcours littéraires en mosaïques. *Revue indépendante*, 338, 18–21.

Chapouthier, G. (2018). *The Mosaic Theory of Natural Complexity: A scientific and philosophical approach* [Online]. Éditions des maisons des sciences de l'homme associées, La Plaine-Saint-Denis. Available at: https://books. openedition.org/emsha/200.

Chapouthier, G. (2018a). Les racines de la complexité en mosaïque. *Revue philosophique de la France et de l'étranger*, 1, 3–10.

Chapouthier, G. (2018b). La mosaïque des traits culturels. *Noema*, XVII, 61–68.

De Waal, F. (1989). *Peace Making among Primates.* Harvard University Press, Cambridge.

Dupont, J.C. (2018). Mosaïque et intégration : éléments historiques et épistémologiques. *Revue philosophique de la France et de l'étranger*, 1, 43–52.

Forest, D. (2018). La complexité en mosaïque à l'épreuve du scepticisme. *Revue philosophique de la France et de l'étranger*, 1, 53–60.

Hammerschmidt, K., Rose, C.J., Kerr, B., Rainey, P.B. (2014). Life cycles, fitness decoupling and the evolution of multicellularity. *Nature*, 515, 75–79.

Heiser, M. (2012). SMiLE: Brian Wilson's musical mosaic. *Journal on the Art of Record Production*, 7.

Hombert, J.M., Lenclud, G. (2014). *Comment le langage est venu à l'Homme ?* Fayard, Paris.

Kaplan, F. (2013). Complexité et organisation hiérarchiques des objets techniques. In *L'homme, l'animal et la machine – Perpétuelles redéfinitions*, Chapouthier, G., Kaplan, F. (eds). CNRS, Paris.

Laming, D. (2015). La ville, une mosaïque complexe et vivante. In *Mondes mosaïques (Des astres à la ville, du vivant aux robots*), Audouze, J., Chapouthier, G., Laming, D., Oudeyer, P.Y. (eds). CNRS, Paris.

Lestel, D. (2001). *Les origines animales de la culture.* Flammarion, Paris.

Libby, E., Rainey, P.B. (2013). A conceptual framework for the evolutionary origins of multicellularity [Online]. *Physical Biology*, 10. Available at: http://iopscience.iop.org/article/10.1088/1478-3975/10/3/035001/pdf.

Michod, R.E. (1999). *Darwinian Dynamics: Evolutionary Transitions in Fitness and Individuality.* Princeton University Press, Princeton.

Michod, R.E. (2009). Socialité et sexe: coopération et conflit, des molécules aux sociétés. In *Aux Origines du sexe*, Civard-Racinais, A., Gouyon, P.H. (eds). Fayard, Paris.

Monod, J. (1970). *Le hasard et la nécessité.* Le Seuil, Paris.

Monod, J. (2003). In *Anthologie du haïku en France*, Antonini, J. (ed.). Aléas Lyon.

Nurock, V. (2018). La morale en mosaïque. *Revue philosophique de la France et de l'étranger*, 1, 23–30.

Ohno, S. (1970). *Evolution by Gene Duplication.* Springer, Berlin.

Oudeyer, P.Y. (2015). Quand les machines créent leur propre complexité. In *Mondes mosaïques (Des astres à la ville, du vivant aux robots*), Audouze, J., Chapouthier, G., Laming, D., Oudeyer, P.Y. (eds). CNRS, Paris.

Robert, S. (2018). L'architecture fonctionnelle du langage et le modèle de la complexité en mosaïque. *Revue philosophique de la France et de l'étranger*, 1, 11–22.

Robert, S., Chapouthier, G. (2013). The mosaic of language. In *Les Origines du langage et des langues*, Fracchiolla B. (ed.). L'Harmattan, Paris.

Saniotis, A. (2013). Evolutionary complexity in relation to Georges Chapouthier's mosaic biological structures [Online]. *Biocosmology-Neo-Aristotelism*, 3/2, 334–340. Available at: http://en.biocosmology.ru/electronic-journal-biocosmology---neo-aristotelism.

von Bertalanffy, K.L. (1968). *General System Theory: Foundations, Development, Applications*. George Braziller, New York.

Wang, W.S.Y. (2007). The language mosaic and its biological bases. *Journal of Bio-Education*, 2/1.

13

Converging Technologies or Paradoxes of Power

Jean-Michel BESNIER
Sorbonne University, Paris, France

13.1. Introduction

Today, philosophers approach representations relating to the phenomena of morphogenesis with a reading grid that the life sciences have largely imposed on them: instead of the traditional opposition between matter (ΰὰ) and form (μορφή or εου ή), they substitute that of form (μορφή) and power (δύναμις or ἐu ς ιςα). Recent developments in biotechnology have made this substitution somewhat more necessary, without, in the eyes of philosophers, dispelling the problems put forward by the disqualification of the old conception, which demanded that we think of life as a substantial form, whose integrity is ontologically based. In bioethical debates, and in an archaic if not conservative way, it has become commonplace to frown upon anyone involved with defending the cause of life, as it was conceived and as early as Aristotle, as the property of matter organized to function as a self-regulating organism, an open system capable of resisting environmental aggression. It is as if progress in the understanding and mastery of the living made it necessary to get rid of all metaphysical substantialism, and to replace it with a dynamic conception of vital phenomena, a conception that could integrate plasticity and malleability among its essential attributes. Thus, the argument developed today by the proponents of anthropotechnics – that is to say, technological devices applied to modify humans – always follows the same line: "You refuse the manipulations of the living permitted by biotechnologies, whereas life itself is the result of a continuous process of individuation; you claim to defend the cause of life, even though you oppose the

The Explosion of Life Forms,
coordinated by Georges CHAPOUTHIER and Marie-Christine MAUREL. © ISTE Ltd 2020.

evolution that continually results from it; if you ignore this evolution, it is because you are a victim of the fixist prejudices that are the mark of religious minds ...." In short, it now seems obvious that it is time to share the following thesis with biologists, trained in the truth of Darwinism: in the world of life, nothing subsists, everything evolves and offers itself to mutations and transformations. Ignoring this evidence of the prevalence of evolutionary dynamics over substantial form would expose us to vain difficulties. The philosopher François Dagognet explained the Westerners' reluctance to encourage body manipulations such as organ transplants: "Aristotle's heirs, he used to say, we always think of life by reference to the organs, so that any manipulation seems to us to be detrimental to the integrity of the organism; if we thought like the Orientals, we would think of life from living tissue, so that organ transplants or transfers would seem to us to be within the norm of possibilities of the living, as the very expression of its potentialities" (Dagognet 1988). The opposition is undoubtedly schematic. It nevertheless suggests that the debate between techno-progressives and bioconservatives (according to the terms of an opposition invented by a few transhumanists today) involves an opposition between a substantialist and a processualist frame of reference (Alexandre and Besnier 2016). The issue at stake in this debate therefore revolves around the notion of form: is life contained within the limits of a form or does it consist of making the provisions of a matter prevail over all forms? Another philosopher, Gilbert Simondon, is sometimes invoked to resolve this alternative, because he considered that our spontaneous conceptions were biased by a closed metaphysics: according to him, this metaphysics is fundamentally hylemorphic, in the sense that it claims to elucidate the mystery of the association of matter and form from which all reality is supposed to derive (Simondon 2005). How does this fundamental dualism appear as the problem of problems, not just for philosophers, but also for scientists called upon to account for physical and biological phenomena? Answer: from an inveterate and inconsistent substantialism. Inconsistent, indeed, because if all reality is the result of the composition of matter and form, it would be difficult to explain, in the hylemorphic couple, that matter is not already form, nor that form is not already matter, allowing each to be made available to the other. Simondon's philosophy was precisely to escape from this difficulty and to undertake understanding the very process by which individuation takes place – that is to say, taking the form of an entity – which stabilizes a coupling without origin or end, for a time. Can we think of this process as the first, and no longer as a derivative of an initial and ontologically unexplained dualism? This is what Simondon believed, without having anticipated that the transhumanists would announce the forthcoming triumph of "information", which would put an end to the limitations imposed by forms (Kurzweil 2005). Such is the context that now gives the concept of power a visionary scope, clearly opposed to the supposed inertia of the advocates of the natural order: form presupposes a stability that challenges power to assume itself, to overrun it and restore the creative chaos from which it would have emerged.

13.2. Might, domination, power

It is therefore not insignificant that the notion of power appears as a key word in the argumentation of technoscientific programs as much as in the rhetoric of geopolitical discourses. Power – and not might or domination: this lexical choice calls for an examination that must be drafted here, if only to contribute to the understanding of our time devoted to globalization, the self-propulsion of technology, as well as to a certain axiological perplexity. The term "power" is nowadays attributed to mega-machines, often grouped together under the name "GAFAM" (Google, Apple, Facebook, Amazon, Microsoft), to these companies to which transhumanists are attached, as well as to the spearhead of humanity without weakness, which they intend to make.

When talking about power, authority is usually brought up first. Indeed, power presupposes an authority commonly defined as the act of one who is entitled to say "I", as the manifestation of the one who shows himself to be the "author" and responsible for the series of decisions he takes. Dominance, on the other hand, presupposes mastery acquired within a given balance of power, while power, it seems, above all, denotes an appetite for growth that seeks to extend and be translated in space and time, such as on the scale of a given territory.

These three terms – might, domination and power – therefore certainly connote something in common; a certain hold on beings and things, but they do not imply subjectivity, decision and consciousness to act to the same degree. In the semantic trajectory from might to power, a logical pathway that increasingly erases subjectivity, decision or awareness of action, can be identified. In fact, power is usually described as having an indeterminate agent as its support, as well as a territory with fluctuating contours, for example, a nation more or less circumscribed within limits, or a consortium, or a multinational. Power thus appears as a process that has got rid of the formal conditions necessary for subjectivity, and which manifests itself as the effect of an anonymous force, of an amplifying mechanism. In order to understand the meaning of power today, one must read Elias Canetti's *Mass and Power* (Canetti 1966). Indeed, Canetti showed how power is contemporary with the advent of the masses. According to him, it is now that which is indistinct, neutral, and transindividual – in this sense, that which is informal and non-subjective, in other words: the mass that best expresses power as it is hypostasized in the modern world.

Even better than Canetti, and before him, it was Ernst Jünger who seemed to have the most impact in understanding the deep meaning of the term "power", when it was used to characterize the technical world and the impact of machines on our representations and lifestyles. In the 1930s, Ernst Jünger wrote a book, which took almost 50 years to translate into French, entitled *The Worker* (Jünger 1932). In this book, which draws lessons from the First World War, he describes the advent of Power, written with a capital P, as the figure of a new type of humanity, which emerged during the First World War. The worker he is talking about is the man of the masses, who is at odds with the bourgeois space. A remnant of a twilight world at the end of the 19th Century, the bourgeois is depicted in a somewhat schematic way as tense on security and haunted by the act of calculating reason. But what happens to this bourgeois? With the First World War, he is on the verge of disappearing. On this subject, Jünger meets Stephan Zweig again, and the formidable book that is *The World of Yesterday* (Zweig 1941). In the first chapter of this book, Zweig sketches the landscape of the secure world that was the bourgeois world, where heritage was passed on, where property was obsessively protected, where forms were essential. What Jünger and Zweig describe, together with other 20th Century writers, is the disappearance, the ruin, of this well-structured world and the advent of the worker involved in a now-risky environment, where that which is irrational is displayed, where elementary powers and impulses, so recognizable in the *pathos* of the time, emerge: in gambling, in adventure, in hate, in love, in short, in everything that is unmotivated and uncontrolled. What Jünger suggests, and what seems so right, is that, without a doubt, this world of the worker is the remake of what the romantic space once criticized the bourgeois world of bringing together, but a romantic space now transformed and constrained by a context increasingly perceived as absurd, given over to that which is unpredictable, incalculable, in short, overwhelmed by the circumstances conducive to the revolutions that would shake the century. For the Worker is indeed the man of these revolutions, or rather, one should say, the man involved in deterritorialization who contributes to the revolutions. An emblematic moment, the First World War thus represents the space in which the elemental powers would emerge, from which the aspiration for a new order would emerge, which Jünger points out: "*in this new order, calculations would not aim to eliminate that which is dangerous* – that is to say, calculating reason can be intended for something other than ensuring security (JMB) – *but it is a world that would have been generated by a new alliance of life and danger.*"

13.3. Life, technique, power

All these themes seem very suggestive at a time when our society is described as a "risk society", a time where there is a lot of talk about the risk tolerance of which we are capable. Our time remains concerned with the description of a new order, symptomatic of a will to power, capable of liberating the elementary and making it the instrument of collective mobilization. This will to power is indeed a will to liberate what was frozen in forms, to liberate the original, the elementary, and perhaps to contribute to the reconfiguration of a new order. What Jünger showed very well in the 1930s was the part played by technology in this reconfiguration. It was indeed technique that helped to dissolve the old states and collaborated in the First World War, in absolutely gigantic and traumatic proportions. It is technology that, in a way, is redistributing the cards and rearranging the structures that defined the world before. Jünger, for example, talks about the opposition of the city and countryside, or the importance given to the value of gold, and he shows how all of this has been shaken by the rise of technology, to the point where the latter appears to be like power, allowing the figure of the worker to mobilize the world. Lastly, it is technique that carries the life and danger alliance, which the bourgeois world managed to stifle.

Today, more than ever, we are forced to think of technique according to the model, evoked here perhaps too literally, of "a process without subject", whose "authority" has faded and whose power is now embodied in self-regulating machines. Machines that cybernetics undertook to build in the late 1940s and that have been actualized in today's learning machines, artificial intelligence's deep learning. It must be said that, at the end of his life, Heidegger was right to consider the appearance of cybernetics as the major event of our time (Heidegger 1976). Indeed, he saw it as the very emblem of technical power. He believed that cybernetics was establishing a world in which technical process was increasingly being seen as fatal and inaccessible to individual initiative. With and after Heidegger, this technical process would be rightly presented more and more as a kind of autonomous, endogenously functioning mechanism over which we lose control (Ellul 2004). For this reason – and this is the main point to be made – power has easily found its best metaphorical translation in biology. Indeed, this concept of power is easily illustrated by Bichat's description of life in the early 19th Century as a spontaneously self-organizing device: "*life is the totality of the forces that oppose death.*" The immunological system was being discovered at that time and this definition of life has received a whole posterity among philosophers. Today, it still seems very relevant to qualify power as the concept that is used today. Before Jünger, it was Nietzsche, of course, who best expressed it by defining life as "will to power". Not "will to might", since power presupposes an "I" that is responsible, but "Life" as a process that wants growth. Life wants growth in all its forms, and in this sense, it is a blind process; it is pure "Will of will". Before Nietzsche, Schopenhauer also formulated it in a striking manner by revealing the existence of a "universal will to live" in the depths of every being, of which we are mere avatars, as if there was a kind of appetite for growth on a cosmic scale, one that was inclined to incarnate in the limited individuals that we are. Of course, we can always give ourselves the illusion of being authors, of having authority over ourselves, but in reality, we are overwhelmed, traversed by a process that is the very process of the "Will to power". From that, to the fact that everything that can be described by a propensity to develop without reflection, without regulation, without finality, is in the end considered as the expression of power, there is only one step that seems to have been taken by the transhumanist prophecies.

It is to this thoughtless, unregulated development, to this unlimited potentiation, that activities and action programs designed to stimulate technological innovations of all kinds are increasingly subjected (Baindbridge and Roco 2003). This potentiation of action programs, whether it be industrial or technoscientific research programs, updates and illustrates the desire for power in this manner, just as philosophers have sought to describe and identify it in mental representations and concepts. And it is clear, in this sense, that technology is indeed the place where power is expressed. Not in the sense, that has long been explained, of highlighting the act of moving something from that which is merely virtual to that which is real. The explanation goes back to Aristotle, who offered a first answer to the philosophical question: what is technique? The gesture by which something that is only in power is actualized. Thus, the skill of a sculptor who forms something out of marble, thus illustrating the actualization of a being in power – the marble statue of Hermes – through the application of a technical gesture that brings out this form (Aristotle 384 BCE). This sense is no longer today's definition of technique. Rather, we would say that it is the set of means and devices that make it possible to multiply that which is possible, to virtualize realities and thus, in this sense, to justify these possibilities in order to prove it to be superior to that which is real. Today, all technoscientists would admit that it is in these terms that they think of the relationship between what is potential and what is real. There is no longer a subordination of what is virtual to what is real, but on the contrary, an increased opening of what is real to what is virtual (Besnier 2006). Technique is thus now the very element that makes it possible to maximize what is possible and to make our world a particular and contingent reality, threatened by inertia and obliged to allow itself to be potentiated in order to survive. The argument of the augmented human finds its spring here. This kind of acceleration of the process of potentiation is therefore what qualifies power. It is for this reason that we associate "power" and "technique" without further thought. The latter is quite naturally described according to an asymptotic law of increase, which reveals potentiation as the norm of all innovation. When we speak of it in this way, we never fail to invoke Moore's famous law, which is supposed to measure the exponential rate of technological improvement. This law in fact formulates the continuous measurement of the growth of technologized societies and signifies the destiny of power to which we are irresistibly committed in a very emblematic way, if we are to believe the thurifiers of transhumanism. And by "destiny", we specifically mean the deprivation of initiative and subjectivity, the relativization of decision, etc.

13.4. "Technological arrogance"

It is not surprising that, from this perspective, power has become a pretext for an unmitigated disaster. As early as Plato's contemporary Greeks, philosophers associated the resources of technique with social and moral disasters. Since then, the theme of the technological hubris seems to have been confirmed, including among some technologists. In 2000, Bill Joy, a very high-flying American engineer (who notably invented the Java language, which we use every day), released a text entitled *Why the future doesn't need us*, in which he wrote the following, which has been relayed countless times since: "As a result, by the year 2030 – all the techno-prophets or all those who castigate techno-prophets, consider 2030 to be the year where there's going to be a break (JMB) – we should be able to produce quantities of units a million times more powerful than today's computers. The combination of this tremendous computing power with advances in manipulation in the physical sciences on the one hand, and recent crucial discoveries in genetics on the other, will unleash a wave of phenomenal transformative power" (Joy 2000). It is as if a mechanism has been put in place to impose a regime of control in developed societies, proving, in the process, that power based on technology is an endogenous process, as far as the organism as biology understands it.

Even when associated with the demiurgic power that is often demonized, technoscientific power is still sometimes, in spite of everything, conceived as the manifestation of the superiority of mind over matter. There is therefore still a temperate way of thinking of technoscientific power as a power of transformation, where the mind that imposes its forms is still at the helm. However, from wisely programmed transformation to the intoxication of "anything is possible", the shift is imaginable. In his book, Bill Joy rightly quotes the physicist Freeman Dyson, who considered that, in the context of the arms race, the scientist at the helm of the nuclear industry remains a mind that runs things, but in such a fascinating way that he might be willing to let himself be overwhelmed. To quote Freeman Dyson: "I have personally felt this fascination for atomic weapons. For a researcher, their power of attraction is irresistible. To feel this energy that sets the stars ablaze, there at your fingertips, to release it and to feel like the master of the world, to perform miracles, to catapult tons of rocks into space, by the millions, this is something that gives you the illusion of limitless power and, in a way, where all our ills come from.

This thing is called 'technological arrogance'. It is stronger than you. You realize the unbelievable power of the mind and it is irresistible." This "technological arrogance" even allows itself to be "naturalized" and ends up translating the fact of biological life itself, like Spinoza's *conatus* describing the natural effort of the living to persevere their own being. As a result, it encompasses and overflows what human beings may want.

Schopenhauer expressed this overflow quite well when he said: "Of course, man can do what he wants, but he cannot want what he wants." And this is precisely the tragedy of power, which distinguishes it from the might and domination that we mentioned in the introduction. Power ultimately escapes the mind, that is to say, human control. The tragedy of power paradoxically justifies the feeling of powerlessness that is increasingly described as the contemporary pathology of men living in technologized societies. A 20th Century German philosopher, Günter Anders, accurately described this feeling of powerlessness generated in men by technological power. In his book *The Obsolescence of Man* (1956), he said that deep down we all suffer from "a Promethean shame of being oneself". He described the feeling of dispossession we feel for the machines that surround us and that we feel we are continually chasing, because they are infinitely more efficient, more effective and more durable than we are. This Promethean shame of being oneself, revealing the feeling of powerlessness, is comparable to this tragedy of power at odds with the might and control that could still structure it, and it rightly condemns us to the dispossession of the initiative.

13.5. Technological convergence and singularity

When contemporary ethics says that we must seek to control our own mastery and that we must fight against technoscientific excess, it is sometimes argued that it is a good thing because it lets us think that we are still in control. In fact, we are perhaps no longer in the position of having to limit our capabilities, but rather of having to face the effects of the control of which the almost endogenous evolution of techniques has forced upon us (Dupuy 2004). Technological activity is a biological process that is self-sustaining and self-feeding, as many paleoanthropologists have explained. André Leroi-Gourhan thus presents it by trying to deduce it from the generic attitude of men, capable of producing tools that are like their prostheses. In Werner Heisenberg's book *Nature in Contemporary Physics*, this understanding of the production of tools has also been described very well, openly stating that "technique is a biological process which, by its very nature, is removed from human control". This idea has almost become commonplace, which supports the axiom that has been attached to the physicist Gabor: "Everything that is technically feasible will be achieved, whatever the ethical costs." This rule reminds us of how much the power we are questioning today refers to elementary life, not allowing itself to be stopped for long by prohibitions and social norms. The age of technology is therefore the age of fundamental disinhibition. How many examples of today's prohibitions currently oppose technical achievements that everyone knows will transgress?

What has been called "technological convergence", since a famous report submitted to the American government in 2003, and which is summarized by the acronym NBIC (Nanotechnology-Biotechnology-Information-Cognition), makes it clear that the line that still seems to separate technology and life is becoming increasingly blurred (Ganascia 2017). In this "technological convergence", we can increasingly see the imminence of a conjunction between what is artifact and what is biological. The example of artificial intelligence, in other words, the announced triumph of non-biological intelligence, already makes it necessary to confront the prospect opened up by NBICs in order to reduce humans to pure algorithmic mechanisms. In the same way, biotechnologies are obviously an attempt to manufacture humans, in other words, to replace births resulting from sexual reproduction with the artifactual production by cloning. Nanotechnology, working on a scale that does not know the distinction between the living and the inert, also promises to produce self-organizing devices, "molecular assemblers", that will make it possible to infiltrate the body to repair DNA and, in any case, to suppress disease or aging (Drexler 1990). In short, each of the components of technological convergence is already seeking, in a way, to develop like a living process that seeks to reshape, augment, and even replace the life of which it is the initial product. It is all these attempts that fascinate the transhumanist movements surrounding laboratories that are dedicated to technological convergence and attached to the idea that technologies are in the process of making the advent of a "successor to man" or a "singularity" possible. In a way, some transhumanists claim that we are introducing variations into the world that are already subject to selection by the market, similar to those that nature itself has always welcomed, and which, after mutation, lead to the emergence of "something new" – a post-human, if you like – that would appear as the embodiment of the power that comes from the mastery to which we are dedicated (Besnier 2009).

These transhumanist movements have momentum, even though they seem to be a thing of fantasy, even responsible for a few impostures. Nevertheless, they are gaining more and more support in the United States and elsewhere, and political and even academic circles are becoming increasingly attentive to their predictions. Why is this? To a large extent, because they endorse the values that the capitalist world claims and flatter a libertarian inclination towards a defamation thrown on politics:

the values of unlimited growth and competitiveness considered as sources of all prosperity, the growing rejection of centralizing institutions (Damour 2018). Naturally, these transhumanist movements echo fascination for technological innovation, considered as the very engine of wealth. Indeed, the motto of technologized societies is to encourage innovation at all costs. To be convinced of this, one only has to read the European Community's Horizon 2020 Program, which replaced the Lisbon Strategy, and which prescribes the technoscientific orientations of European policies. Innovation is the buzzword, and in the calls for projects, it replaces the reference to scientific research itself. We can understand the concerns of today's researchers, who complain about being forced into the short-term duration of programs, devoid of any visionary inspiration and oriented, on the contrary, towards the production of objects or innovations that are preferably "disruptive": the rush forward and incantation have become the rule and hype announcements have become the food of choice for contemporary research policies.

13.6. Innovation, research, invention

Making innovation the exclusive objective of research is increasingly becoming a kind of perversion. From this point of view, some great historians of technology, such as Bertrand Gille and today's Bernard Stiegler, have been right in suspecting that the slogan of innovation is a deviation or even a subversion of the ambitions of research. Twenty years ago, Gille wrote the following: "Innovation is no longer what results from an invention, it is a global process aimed at giving rise to invention, and has emergence as its program." He also commented on the sort of inversion, where innovation comes before invention: "Previously, invention had to wait to be applied until the technical, economic, social, etc., conditions became favorable. Innovation followed. Now, it is the desire for innovation that drives invention. It is on this inversion that research and development (R&D), and what can now be called "technoscience", are based. There are many pieces of work that demonstrate this progressive crushing of research on innovation, revealing how technical invention and scientific discovery are being confused, mainly because of the reduction in the time taken to transfer the discovery from physical phenomena to an invention, and then to innovation. Some data are produced to illustrate this transfer: while it took, for example, 102 years for knowledge related to optics to eventually produce photography and 56 years for knowledge about electromagnetic waves to produce the telephone, it took only 12 years for television, 14 years for radar, 6 years for the uranium bomb, 5 years for the transistor, etc. We must therefore record this kind of knowledge, which is obviously innovation's justification in becoming the very breeding ground for invention and scientific research.

As it is claimed as a manifestation of power by the developed countries, innovation risks killing science that was seen, perhaps naively, as the result of a selfless love of knowledge. The proponents of slow science are now ready to make a deal with the "objectors of growth" or the supporters of voluntary simplicity, as they are multiplying today, particularly in Canada and against California's GAFAM or China's BATX (Baidu-Alibaba-Tercent-Xiaomi).

The fight against such economic and industrial powers is undoubtedly unequal, perhaps devoted to symbolism and speech. Nevertheless, the demoralization of researchers who do not want to bow down to the will of the dominant power is a serious matter, which could lead to a growing demobilization and, ultimately, to the triumph of the sole objectives of ergonomics and design in technoscience, with which innovation is mainly concerned. Perhaps an equal power could turn against it and reveal how a civilization dies.

13.7. Conclusion

Attachment to forms – to rituals and procedures – is enshrined in the spirit of the laws that guarantee societies' stability. On the other hand, the fascination with power, especially when it seizes might, never brings about forms. Will our time, then, be proof of "deformation"? The impact of NBICs on the imaginations of a growing number of individuals is not without consequence: it expresses the technological brutality to which we are accustomed, when they demand ever greater speed and efficiency, when they brand us of being impatient and inattentive. Of course, the cult of forms is not decreed. At the very least, it is worth recalling that it has long been associated with the ideals of the "good life". If too much formalism kills life, the absence of form equals violence and death. With all due respect to the technoprophets of transhumanism, their fantasy of immortality, fueled by virtual technologies or the frightening prospects of genetic engineering, forces the living to dispute that the vulnerability of mortals is not a symptom of impotence, but the guarantee of this incessant invention of forms, which we call "culture".

13.8. References

Anders, G. (2002). *L'obsolescence de l'homme. Sur l'âme à l'époque de la deuxième révolution industrielle*. Les éditions de l'Encyclopédie des nuisances, Paris.

Alexandre, L., Besnier, J.-M. (2016). *Les robots font-ils l'amour? Le transhumanisme en 12 questions*. Dunod, Paris.

Aristote (2000). *Métaphysique*, IV, 6. Vrin, Paris.

Baindbridge, W.S., Roco, M.C. (2003). Converging Technologies for Improving Human Performance. Report, National Science Foundation, Dordrecht.

Besnier, J.-M. (2006). *La croisée des sciences. Questions d'un philosophe*. Le Seuil, Paris.

Besnier, J.-M. (2009). *Demain les posthumains. Le futur a-t-il encore besoin de nous?* Hachette, Paris.

Canetti, E. (1966). *Masse et puissance*. Gallimard, Paris.

Dagognet, F. (1988). *La maîtrise du vivant: fondements d'une biopolitique*. Hachette Littératures, Paris.

Damour, F. (2018). Transhumanisme et biocapitalisme. *Raison Présente*, 205.

Drexler, É. (2005). *Engines of Creation. The Coming Era of Nanotechnology*. Anchor Books, Doubleday.

Dupuy, J.-P. (2004). *Pour un catastrophisme éclairé. Quand l'impossible est certain*. Le Seuil, Paris.

Ellul, J. (2004). *Le système technicien*. Le Cherche Midi, Paris.

Ganascia, J.-G. (2017). *Le mythe de la singularité. Faut-il craindre l'intelligence artificielle?* Le Seuil, Paris.

Heidegger, M. (1976). *Réponses et questions sur l'histoire et la politique*. Mercure de France, Paris.

Joy, B. (2000). Why the future doesn't need us [Online]. Available at: www.wired.com.

Jünger, E. (1932). *Der Arbeiter*. Hanseatische Verlags-Anstalt, Hamburg (English translation: *The Worker*, Northwestern University Press, 2017).

Kurzweil, R. (2005). *The Singularity is Near: When Humans Transcend Biology*. Penguin, New York.

Simondon, G. (2005). *L'individuation à la lumière des notions de forme et d'information*. Million, Grenoble.

Zweig, S. (1996). *Le monde d'hier: souvenirs d'un Européen*. Le Livre de Poche, Paris.

List of Authors

Jean-Michel BESNIER
Sorbonne University
Paris
France

Dalila BOVET
University of Paris Nanterre
France

Georges CHAPOUTHIER
ICM-CNRS
Paris
France

Laurence DEVILLERS
LIMSI-CNRS
Sorbonne University
Paris
France

Avelina ESPINOSA
Roger Williams University
Bristol
United States

Didier FASS
LORIA-CNRS
University of Lorraine
Nancy
France

Jean-Pierre GASC
National Museum of Natural History
Paris
France

Florian JABBOUR
National Museum of Natural History
Paris
France

Robert JAFFARD
INCIA-CNRS
University of Bordeaux
France

Andreas LOSCH
Faculty of Theology
University of Bern
Switzerland

Guilhem MANSION
Freelance biologist
Zurich
Switzerland

Marie-Christine MAUREL
ISYEB-CNRS-MNHN
Sorbonne University
Paris
France

Guillermo PAZ-Y-MIÑO-C
Roger Williams University
Bristol
United States

Jonathan B. WEITZMAN
CNRS
University of Paris
France

Index

E, F

G, H

I, J

K, L

M, N

O, P

Printed and bound by CPI Group (UK) Ltd, Croydon, CR0 4YY

13/09/2026

14961043-0001